Icelandic Rocks
and Minerals

KRISTJÁN SÆMUNDSSON   EINAR GUNNLAUGSSON

ICELANDIC ROCKS AND MINERALS

PHOTOGRAPHS: GRÉTAR EIRÍKSSON

ENGLISH TRANSLATION: ANNA YATES

Mál og menning

Text © Einar Gunnlaugsson and Kristján Sæmundsson
English translation © Anna Yates
Photographs © Grétar Eiríksson, except © Sigurður Sveinn Jónsson (SSJo)
101 top left, 103 bottom left, 107 top, 113 top right, 115 top right, 117 top, 133 top,
137 bottom left, 156 top left, 170, 177 bottom, 179 top, 191 top, 195 top.

First published in Icelandic by Mál og menning © Reykjavík 1999.
This edition published in 2002 by Mál og menning ©, an imprint of Edda Media & Publishing,
Reykjavík, Iceland.

Cover design: Næst
Cover photograph: Grétar Eiríksson
Diagrams: Næst, Helga Sveinbjörnsdóttir and Einar Gunnlaugsson
Maps: Jean-Pierre Biard

Printed in Denmark by Nørhaven Book

ISBN 9979-3-2199-7

CONTENTS

Introduction

The rocks of Iceland are less varied than those of many other countries. The average person knows little of rocks and minerals, but many notice unusual rocks they come across, and may even take them home. Hitherto, people with an interest in rocks have had to make do with foreign books which cover a vastly greater range of rocks and minerals than are found in Iceland, as no Icelandic handbook has existed. This book, which includes only Icelandic rocks and minerals, is intended to meet this need. The book includes descriptions of the principal minerals and rocks found in Iceland, and the main variants. The book is not exhaustive, however, as most minerals have been omitted that cannot be seen with a good magnifying glass. The book is written for the layman who wishes to learn to identify the most accessible and noticeable rocks and minerals that occur in Iceland. Those who wish to learn more can easily do so with the help of specialised handbooks and other aids.

Anyone who wishes to learn about Icelandic rocks must understand Iceland's geological structure. Hence the geology of Iceland is described in a brief overview which explains the principal features of its structure and the main geological formations. Secondary minerals are most sought after. Their formation is contingent on various factors which must be understood if one goes in search of some special mineral assemblages. So it is worth reading the introductory section on altered rock and amygdules. Collectors are generally less interested in rocks, but they show considerable variety and are discussed here in some detail.

Minerals are classified in principle by chemical composition, while in two chapters of the book they are classified by the conditions of formation, which was regarded as a more suitable framework. Many minerals are named after places and people. In some cases, special Icelandic names and terms are mentioned.

The photographs in the book are intended to show rocks and minerals in their commonest form. Variants are included, but the aim is to illustrate identifying features. There are few photographs of eye-catching mineral specimens, and those few are illustrative of the exterior identifying features. At the end of the book is a table that summarises the main properties of the minerals.

Locations where minerals are found are stated only in general terms. If one knows exactly where to look, the search for rocks is much less fun. In general, the region or fjord is named, or the mountain in the case of rare minerals. The East Fjords are known for their minerals. Of the famous locations there, two – Helgustaðir and Teigarhorn – are protected sites. But there are rich hunting-grounds for minerals in many other areas, both in the north and the west, especially in and around Hvalfjörður. High-temperature minerals are found mostly in Austur-Skaftafellssýsla in the southeast, and the same applies to ore minerals, as major intrusions create favourable conditions there. When searching for minerals and collecting them, one should always speak to the landowner first. Few make any objection, as mineral-collectors rarely do any damage.

People have always admired crystals. Before the principles of crystal formation were understood, they were often believed to possess supernatural properties. Icelandic folklore, for instance, abounds in stories of stones of invisibility, wishing stones, etc. As understanding of their structure and formation increased, these superstitions declined, but they have reappeared in modern times, and claims are made that certain crystals possess curative properties. This is entirely unsupported by science, but this superstition sells well.

Many people have contributed rock specimens to be photographed for this book, especially *Hermann Tönsberg* who has a large and extensive collection of minerals. He was also a source of abundant information on minerals found in Iceland, and the locations where they were found. The *Icelandic Museum of Natural History, Steinaríki Íslands* (Mineral Kingdom) at Akranes, and *Álfasteinn* in Borgarfjörður eystri loaned us many specimens, and the directors of these rock collections gave us much useful information. Specimens were also lent by other people. At the back of the book is a list of the owners of the minerals illustrated in the book.

The *Orkustofnun* gave permission for X-ray analysis of several dozen minerals. This is gratefully acknowledged. *Sigurður Sveinn Jónsson* provided invaluable help in the preparation of this book by carrying out the X-ray analyses. He also took some of the photographs, including those of some of the smallest minerals. His photographs are identified by his initials in the captions.

The earth's crust and the uppermost part of the mantle are known as the lithosphere, below which the mantle is soft and plastic. Convection currents in the mantle move the tectonic plates of the lithosphere away from mid-ocean ridges where basaltic magma flows up from below to be added to the edges of the plates. Iceland is located at the boundary of two such plates. It is one of two subaerial segments of the mid-ocean ridge system, which is 40,000 km in length. Only 1% of the ocean-ridge system is above sea level.

Each of the largest plates extends over a whole continent and half an ocean, thus comprising both continental and oceanic crust. At convergent plate margins, fold mountains form if one or both margins consist of continental crust, while island arcs form if one or both plate margins are of oceanic crust. Ocean crust eventually returns to the mantle, sinking at the plate margins along a dipping thrust or subduction zone. This is where volcanic activity in fold mountains and island arcs originates.

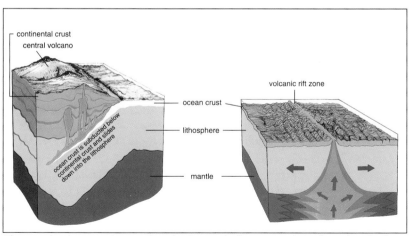

Plate margins. At converging margins volcanism results from magma being formed on the thrust plane, while at divergent margins magma forms where a convection current in the mantle wells up underneath the crust.

Iceland forms a thick swell in the ocean crust, caused by a powerful mantle plume where the upflow of magma is about three times greater than in the ocean ridge to the southwest and north of the island. The mantle plume is hotter and lighter than its surroundings. It has existed since the North Atlantic began to spread apart in the early Tertiary, about 60 million years ago, building up the Greenland-Iceland-Faroes ridge in its wake. The centre of the mantle plume is at present under the Vatnajökull glacier, out of line with the ocean ridges to the north and south. This is because the plume has caused shifting of a segment of the ridge axis to the east in focus with itself. The tectonic plates of the North Atlantic spread about 1 cm per year in opposite directions, which is among the slowest known rates of spreading. East and West Iceland are thus moving apart at a rate of about 2 cm per year (about the height of a man in a human lifetime). The spreading rate is regular at some distance from the rift zone, but there the movement occurs sporadically, after tension has built up for a century or two. Iceland may not actually be increasing in area due to spreading, as marine erosion also acts upon its east and west coasts. The island thus renews itself. If these processes continue to act in their present manner for 20 million years a new Iceland will have been formed, about 400 km across from west to east, in place of the present one.

Iceland's rock strata have been formed in the rift zone, where volcanism has added one lava on top of another, gradually weighing down the substratum. The rock strata were thus gradually tilted towards the rift zone, as more and more lava piled up. At the same time, spreading was at work, moving the plates away from the zone of rifting and volcanism. When no more lava was added to the lava pile, erosive forces took over. Valleys were formed, which were deepened and lengthened by glacial erosion in the late Tertiary and Pleistocene (Ice Age). As the lava pile was eroded, the remaining areas rose in response, forming the mountains of the plateau basalt regions with their fjords and valleys. The more rock was removed by glacial erosion, the higher was the uplift. This is seen most clearly in the southeast. There the depth of erosion at sea level is about 2,000 m below the original surface of the lava pile.

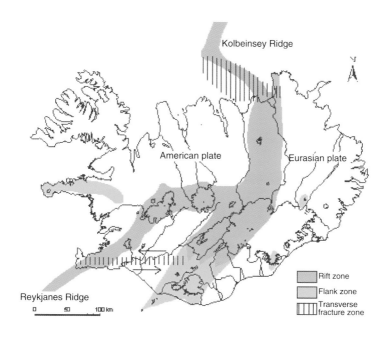

Divergent plate margin in Iceland. The oceanic ridges and the rift zones on land are not aligned. The two are linked by transverse fracture zones, which are leaky as volcanic activity also occurs in them. The flank zones are caused by horizontal strain in older crust outside the rift zone. All Iceland's true stratovolcanoes are found here.

Volcanism and faulting in the rift zone are related to volcanic systems, each of which consists of a central volcano and fissure swarms extending from it to the northeast and southwest, except north of Askja, where the trend is closer to north-south. Central volcanoes produce acidic and intermediate rocks, in addition to basalt. They host high-temperature areas, and shallow magma chambers form in their roots. Magma is extruded from these into curved fissures which slope upwards from them, forming cone sheet swarms. When a magma chamber is emptied a large caldera may form. Eruptions from fissure swarms normally produce only basalt. Their surface expression is a long row of craters. In eroded rock strata, central volcanoes can be identified by rhyolite, strongly altered rock, an abundance of intrusions

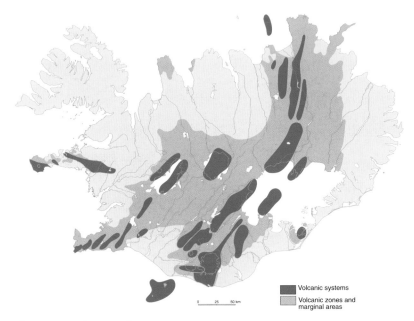

Volcanic systems

Volcanic zones and
marginal areas

0   25   50 km

Volcanic zones and active volcanic systems.

and irregularities in the rock stratigraphy, while fissure swarms may be identified as dike swarms.

Volcanic activity also occurs outside the rift zone, in so-called flank zones, where plate movements have caused horizontal shear and broken up the crust. Magma can rise into the fissures from a greater depth than in the rift zone. This magma is more alkaline than the tholeiite magma of the rift zones. The Snæfellsnes peninsula and the large volcanoes in the South, together with the Westman Islands, belong to the flank zones.

Iceland consists mainly of volcanic rock. Only 5–10% comprises sedimentary rock, but the proportion increases greatly in strata from the end of the Tertiary and the Pleistocene, after glaciation on a regional scale set in. The bulk of the volcanic rock consists of basalt lavas, along with hyaloclastite and pillow lava from glacial periods. The basalt lava layers are of two main types: *pahoehoe* lava (Icelandic *helluhraun*), which is sometimes compound, i.e. composed of numerous flow units, for instance in lava shields, and *aa* or block lava (Icelandic *apalhraun*), usually simple, i.e. consisting of one or few flow units, from crater rows. Pahoehoe lava fields are of two main kinds. On the one hand, they comprise thin lava layers, with an undulating surface and ropy texture, or with a thin surface layer of glassy debris. On the other hand pahoehoe lava may be simple thick flows, with a smooth surface, dotted with deep hollows, but low at the margins. Lava flows of this kind have formed where the surface of the lava has solidified, while molten lava continues to flow beneath, thickening them and disrupting the surface layer. Aa lavas have a jagged surface layer of porous clinker and scoria and are difficult to traverse. The aa lavas have formed from more viscous lava than pahoehoe, which does not release gases so readily. The most jagged of the aa lavas are of intermediate composition, such as the andesite lava erupted by Hekla. Dacite and rhyolite lavas are even more rugged, with angular lava blocks covering the surface.

When lava solidifies and contracts, a network of cracks forms in the lava, growing from top and bottom towards the lava interior. When these cracks form regular five- to seven-sided columns, this is known as columnar basalt (Icelandic *stuðlaberg*). In this case the cooling cracks have formed upwards from the bottom of the lava. This is common where the lava covered sediments, where it cooled quickly, slowing down its rate of flow. In columnar basalt, individual columns may be some metres long, and from 20 cm to two metres in diameter. Entablature lava (Icelandic *kubbaberg*), with irregular small columns, often occurs on top of columnar basalt in the same lava. This has solidified from the surface downward, when water has flowed over hot lava, rapidly cooling it.

Rock strata are often intersected with dikes, where magma has intruded into fissures and solidified. Dikes are generally columnar, with near-horizontal columns. The farther down one delves into the rock strata, the larger proportion of the rock mass consists of dikes. The dikes sometimes occur in swarms, i.e. clusters of near-vertical dikes. At sea level they may comprise some 10% of the rock mass.

Dverghamrar at Síða, South Iceland. Columnar basalt (colonnade) and rapidly chilled *kubbaberg* (entablature) in a Quaternary lava. The twofold division of the flow is due to water flowing over it while hot.

Such dikes are rarely more than 10 metres thick. In eroded central volcanoes, however, they comprise a far larger proportion of the rock mass, with cone sheets and irregular veins criss-crossing their cores. Such dikes are rarely more than two metres thick. In addition to dikes, large intrusions are also known, mostly connected with central volcanoes. The commonest are dolerite, gabbro and granophyre intrusions, which have formed where magma has accumulated at a depth of 1 km or more below the surface. Laccoliths, generally of rhyolite, occur in several locations. These are formed by intrusions between rock layers at a depth of several hundred metres. As a result the overlying layers are pushed upward, forming an updomed roof on the flank of the dome-like intrusions. Volcanic plugs or necks are elongated or pipe-like intrusions which have solidified in the upper part of volcanic conduits. They are either of basalt or rhyolite, often about 100 m in diameter, standing apart as erosional remnants.

Altered Rock and Amygdules

Below the water-table, all cavities and fissures in rock are saturated with water. Fresh volcanic rocks near the surface are generally porous and permeable. Cracks and ground fissures add considerably to the permeability. At greater depths, the permeability decreases due to compression of the rock and settling of mud particles in open spaces. As the rock strata become compacted, heat flow from below leads to alteration of glass and rock. The heat derives from magma, which solidifies in dikes and intrusions in the crust, which is constantly formed in the rift zone.

In Iceland the temperature of the earth's crust rises approximately by 100°C per kilometre of depth at the margins of the rift zone. In the uppermost 500–800 metres of the rift zone, deeply-reaching cold groundwater systems dominate. When the fissures cease to be active they tend to close. As a consequence the permeability decreases, and heat transfer by water circulation gives way to conduction through the rock. As the rock is heated water reacts with the rock, dissolving it, and causing deposition of new alteration minerals in fissures and cavities, in accord with the different heat environment. These secondary minerals in the rock are called amygdules, and the rock thus affected is called altered rock.

Alteration and secondary mineralization are most intense in central volcanoes. Intrusions occur repeatedly in their roots and high-temperature systems become active and persist as long as the volcano. High-temperature areas are found primarily in active central volcanoes in the rift zone. There the temperature gradient is much higher than in the lava areas in their surroundings. The temperature increase approaches the pressure-related boiling point of water.

Alteration of rock and formation of amygdules in the accessible part of the rock strata take place at temperatures of approximately

30–350°C. On the margins of intrusions, temperatures may be higher. Thus a temperature of over 380°C was measured in a borehole near to a dike complex at a depth of 2 km in an Icelandic high-temperature area.

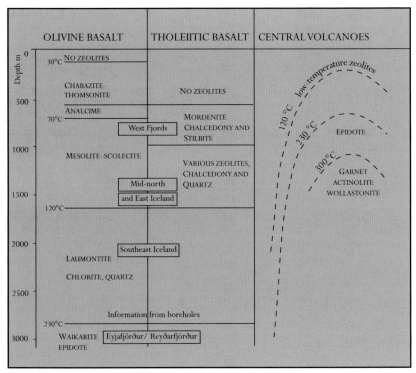

Amygdules in the lava pile. The formation of amygdules depends upon temperature, the type of rock and the composition of the water it contains. Olivine basalt begins to alter at lower temperatures than tholeiite and rocks richer in silica. Zeolite zones are most easily identified in olivine basalt. In tholeiite, quartz minerals and silica-rich zeolites, such as mordenite, stilbite, heulandite and epistilbite are more common. Below the zeolite zones greenschist zone minerals occur with chlorite, epidote and finally actinolite. In central volcanoes, high temperatures occur at shallow depth. There the deeper alteration zones of ordinary lava pile reach highest. The schema indicates how deep erosion has progressed down into the lava pile in the various regions, relative to sea level.

In amygdules, different minerals may grow one on top of another. In this case they occur in the order gismondine – apophyllite – mesolite. From *Hvalfjörður, West Iceland*. Size: *4 x 7.5 cm.*

Common amygdules, such as most zeolites and silica minerals, form at temperatures below 200°C. At higher temperatures, only quartz and two species of zeolites form, along with various high-temperature minerals, mainly greenish chlorite and epidote, which gives a pale-green sheen to the rock at the core of eroded central volcanoes.

As the crust moves away from the rift zone where it is generated, the intrusion activity ceases, and it gradually cools. Amygdules and other alteration minerals, corresponding to the highest temperatures that were reached during the alteration stage, remain in the rock. However, new minerals may continue to form for a time as the crust cools. At lower temperatures different minerals may form on top of the existing ones. Thus one often finds several species of zeolite, which have formed successively in the same cavity. In this way altered rock

17

Stilbite is a common amygdule in fissures in basalt. It lines the walls of the fissure. The space in between is half full of clay (smectite). From *Borgarfjörður, West Iceland*. Size: *width of stilbite lining in fissure 2.5 cm on each side.*

Different minerals in the same amygdule. A coating of chalcedony is nearest to the wall. A layer of clinoptilolite (a variant of heulandite) has formed horizontally in the cavity, after which heulandite grew over the whole. Finally a crystal of sugar calcite formed in the centre. From *Hvalfjörður, West Iceland*. Size: *3 x 3.5 cm.*

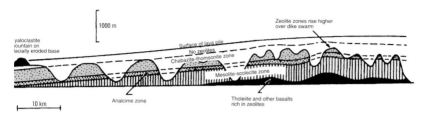

G.P.L. Walker's classic diagram (1960) of the zeolite zones in the east of Iceland. The zeolite zones are identified by amygdules in olivine basalt. The section extends from the East Fjords to upper Jökuldalur.

with a wealth of secondary minerals, once deeply buried, may be exposed in mountain slopes, gullies and sea cliffs.

It has long been known that amygdules are more abundant, more beautiful and more varied, at sea level than near the mountain tops. About 40 years ago a British geologist, George Walker, made a careful study of Iceland's amygdules, and explained the processes involved. He identified the different species of secondary minerals, discovered that they vary according to the lava type, and that they are arranged in more-or-less horizontal zeolite zones. Where dikes are abundant, the zeolite zones are pushed upwards, while each zone spans a larger section of the strata where there are few dikes. Each zone has its characteristic minerals, but the variety of minerals increases the lower one progresses down into the lava pile, in keeping with rising temperatures at the time of formation. Since Walker's study, others have added to the existing knowledge in this field, and sought to define the temperature limits of formation of individual secondary minerals. This knowledge is very useful when boreholes are drilled in geothermal systems, especially high-temperature systems, as it holds a clue to reveal their thermal history.

19

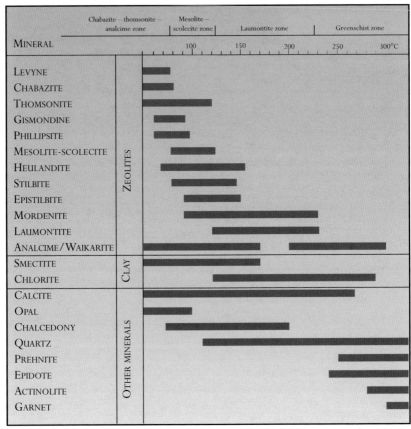

Amygdules in basalt and the temperature range at which they form. The bars show the optimum temperature of formation for several minerals.

The uppermost 200–500 m of the rock strata, depending on the type of lava and the proportion of dikes, is devoid of amygdules. Below this is a zone about 400 m thick where opal, and the zeolites chabazite and thomsonite, are common in olivine basalt. In tholeiitic basalt, only clay amygdules and opal are found in this uppermost zeolite zone. The analcime zone begins where the zeolite analcime is first seen in olivine basalt. This zone also includes all the amygdules found in the chabazite-

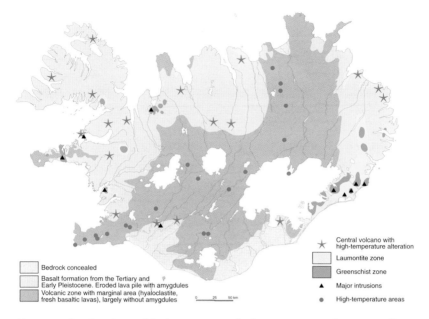

Bedrock concealed

Basalt formation from the Tertiary and Early Pleistocene. Eroded lava pile with amygdules

Volcanic zone with marginal area (hyaloclastite, fresh basaltic lavas), largely without amygdules

0 25 50 km

╬ Central volcano with high-temperature alteration

▢ Laumontite zone

▢ Greenschist zone

▲ Major intrusions

● High-temperature areas

Alteration of rock and amygdule formation in Iceland. In many areas, data are insufficient exactly to define the alteration zones. Eroded central volcanoes show the highest degree of alteration. Within the volcanic zones and their margins, there is abundant alteration only in the high-temperature areas.

thomsonite zone. Near the boundary of these two zeolite zones, phillipsite and its closely related zeolites gismondine and garronite are most commonly found. In the lowest part of the analcime zone, the first zeolites are found in tholeiite, for instance mordenite and stilbite, and chalcedony replaces opal. Below the analcime zone is a zone about 800 m thick, typified by mesolite and scolecite, found mainly in olivine basalt. Mesolite occurs first, as delicate needles, while at greater depth in the zone these zeolites form dense amygdules, the true "ray stone" (*geislasteinn*) of Icelandic nomenclature. Many of the zeolites of the upper zones are also found in the mesolite-scolecite zone, and they even may be at their finest there. In the lower part of the mesolite-

In faults, slickensides are often seen with a coating of amygdules. These are commonly stilbite and ferrous oxides (hematite and goetheite), as seen here. From *Flateyjardalur, North Iceland.* Size: *12 x 8 cm.*

scolecite zone, tholeiitic basalt abounds in zeolites, no less than olivine basalt. The bottom zeolite zone is the laumontite zone, downwards from the first appearance of this mineral. Other zeolites become rarer, leaving only laumontite, often in large quantities. Below the zeolite zones is the epidote zone, where temperatures during formation of amygdules exceeded 230°C. The epidote zone is rarely exposed outside central volcanoes, but in basalt areas boreholes have been drilled down into it, for instance in Eyjafjörður in the north and Reyðarfjörður in the east.

Iceland's lava pile has been eroded down from the original surface to a varying degree. The erosion is greatest at the coast, where only isolated mountains and ridges rise high, interspersed with broad plains, fjords and valley systems. Farther inland where heaths and plateaux are continuous, with shallow valleys, the land is less elevated. This is due to the isostatic equilibrium of the crust, the land rising up highest where most has been eroded. It has been estimated that in the lowlands of the southeast about 2,000 m have been eroded from the lava pile. In the East Fjords and in Eyjafjörður, about 1,500 m have been eroded. Around Húnaflói in the north, west to Breiðafjörður, Snæfellsnes, Mýrar, Borgarfjörður and Hvalfjörður, the erosion amounts to about 1,200 metres, while in the northwestern part of the West Fjords it is only about 1,000 m or less.

In eroded rock strata one often sees fissures of varying width filled with mineral deposits. Hot water has often flowed through these fissures, and the secondary minerals in them may differ from those in the surrounding amygdaloidal rock. Such fissures are excellent hunting-grounds for rock-collectors, as the crystals are often large and regular in form, especially if the fissure is only partially filled and hollow at the centre. Stilbite and calcite are the minerals most commonly found in fissures, while quartz is also common, and laumontite in the most deeply eroded regions. Iceland spar, a variant of calcite, is found only in fissures. The chalk rock that was mined at Mógilsá in Esja, and similar deposits at Þormóðsdalur east of Reykjavík, are fissure fillings. They comprise mainly calcite, while quartz is predominant at Þormóðsdalur. A common form of deposits in fissures is thin flakes or coatings on fault planes sometimes with clear striations. This is generally stilbite, but also hematite and limonite, which give a reddish-brown colour. In hotter surroundings the coatings are quartz and laumontite, with greenish chlorite. These fissures rarely yield minerals sought after by collectors.

MINERALS AND HOW TO IDENTIFY THEM

IDENTIFICATION OF MINERALS

There is a fundamental difference between the concept of a mineral and a rock. A rock normally comprises several minerals, which form for instance when magma solidifies. These are known as rock-forming or primary minerals, as distinct from secondary minerals that are formed when rock is altered, or minerals deposited from hot water in cavities and fissures.

Iceland's minerals all have international names which are used here, while reference is also made to a few specifically Icelandic terms.

Minerals are identified by a number of physical properties, such as crystal form and habit, colour, streak, cleavage, fracture, lustre, transparency and specific gravity. A magnifying glass is a vital tool, and a pointed knife can be used to test the hardness. To test the colour of the streak, an unglazed piece of porcelain or a mortar and pestle is required. Some minerals are soluble in acids, and others, especially metals, can be identified using a flame. In order to make a precise identification, more sophisticated equipment is required, such as a microscope and X-ray diffraction.

CRYSTALS

Several common non-organic chemical compounds are known in nature, which are not recognised as true minerals, as they are not crystalline but amorphous. These are termed mineraloids. These include opal and glass (tachylite, obsidian). Other members of this group are iddingsite (altered olivine), palagonite (altered basalt glass) and chloropheiite (related to chlorite). These mineraloids are included here as variants of those rocks and minerals from which they form under conditions of rapid solidification (glass) and low-temperature alteration.

The crystal form of minerals reflects the internal relationship of the elements of which they are composed, and the way in which they

CRYSTAL SYSTEM	REFERENCE AXES	DIAGRAM	SYMMETRY	EXAMPLES
Cubic	Three axes of equal length, at right-angles to each other.		Three four-fold axes.	garnet, pyrite
Tetragonal	Three axes at right-angles to each other. Two of equal length, the third shorter.		One four-fold axis.	apophyllite
Hexagonal	Three axes in the same plane forming an angle of 120°.		One six-fold axis.	quartz
Trigonal	Three axes of equal length in the same plane forming an angle of 120°.		One three-fold axis.	calcite
Orthorhombic	Three axes of different lengths at right-angles to each other.		Three two-fold axes.	sulphur
Monoclinic	Three axes of different lengths. Two at an oblique angle to each other, the third at right-angles to the plane of the others.		One two-fold axis.	gypsum, epidote, augite
Triclinic	Three axes of different lengths, none at right-angles to another.		No axis.	feldspar

The seven crystal systems and their principal properties.

are arranged in a crystal framework. This is subject to mathematical laws, and hence crystals may be of regular and beautiful shapes if they grow freely. Crystals grow by addition to the faces. If one face grows faster than others, the crystal becomes distorted, but the basic form is maintained, as the angle between the faces is always the same. When crystals are described, the faces of the crystal are placed in relation to imaginary reference axes which all meet at one point at the centre of the crystal. Crystal faces are defined by a system of coordinates.

The internal structure of the crystal is governed by certain laws of symmetry. Take the example of a cubic crystal, i.e. one with six identical faces. Each pair of opposing faces may be linked by an axis passing through the centre of them. This produces three axes of equal length at right-angles to each other, and if the crystal is rotated once, the same form is repeated four times. These are known as four-fold axes. A crystal with three four-fold axes belongs to the cubic system. A crystal shaped like a matchbox has three axes of differing length, at right-angles to each other. When rotated about these axes, the crystal presents the same form only twice. These are thus two-fold axes, and the crystal belongs to the orthorhombic system. Axes of crystals can be six-fold, four-fold, three-fold and two-fold.

All crystals can be classified into seven main systems according to their symmetry. Each system may include many different regular external forms, and there are 32 classes of symmetry. The triclinic system has the lowest symmetry, followed by monoclinic, orthorhombic, trigonal, hexagonal, tetragonal and cubic, which has the highest symmetry. The table above illustrates these seven main systems.

Crystals of the same type often tend to twin, and the twinning is subject to certain laws. Some crystals merge at one face, while others form an intergrowth. Twinned crystals are common both in rock-forming minerals, e.g. feldspar, and in secondary minerals formed by alteration, e.g. zeolites.

HABIT

Habit is the shape of the crystal with reference to the common crystal forms. Crystal habit is described by such terms as: columnar, granular, acicular, fibrous, prismatic, platy, dendritic or amorphous.

COLOUR

When identifying the colour, one must ensure that the colour is not derived from impurities on the surface of the crystal.

STREAK

The colour of the powdered mineral is termed the streak. This may be seen by drawing the mineral across a piece of white unglazed porcelain (hence "streak"), or by grinding it with a mortar and pestle. The streak is a better identifying feature than the actual colour of the mineral.

CLEAVAGE

Some minerals cleave only along one plane, others in two planes or more. The cleavage differs from one plane to another. It can be perfect, distinct, indistinct or none.

FRACTURE

If a crystal breaks accidentally, the fracture, if not a cleavage plane, may be uneven, hackly, conchoidal, splintery or brittle.

LUSTRE

Lustre describes the reflective quality of the crystal's surface. The terms for lustre are self-explanatory: metallic, adamantine (diamond), vitreous, silky, pearly and greasy. Some minerals are without lustre.

TRANSPARENCY

Transparency describes how light passes through the crystal. Some minerals are opaque, while others are transparent, allowing identification of objects when looking through them, or translucent if light passes through them, without objects being identifiable.

SPECIFIC GRAVITY

The specific gravity of minerals varies. Most ore minerals have a specific gravity over 3 to 4, the light-coloured minerals in igneous rock around 2.6 to 2.8, while the dark minerals have a specific gravity over 3. Zeolites, which contain water, generally have specific gravity of 2 to 2.3, and high-temperature minerals around 3. Some minerals or mineral groups can be identified with some precision by weighing them in the hand.

HARDNESS

The hardness of minerals is tested by trying to scratch them with a test mineral, fingernail or knife. The Mohs scale comprises ten degrees of relative hardness. A certain mineral has been taken as characteristic of each degree. These reference minerals will scratch those which are below them on the scale. The softest minerals can easily be scratched with a fingernail. An ordinary knife will scratch minerals of hardness 4–5. Quartz easily scratches glass and feldspar, while the feldspar will slightly scratch glass. The following minerals are used as reference for testing hardness.

HARDNESS SCALE

Talcum	marks fingertip	hardness 1
Gypsum	scratched by fingernail	hardness 2
Calcite		hardness 3
Fluorite	scratched by knife	hardness 4
Apatite		hardness 5
Feldspar		hardness 6
Quartz	scratches glass	hardness 7
Topaz		hardness 8
Corundum	scratches quartz	hardness 9
Diamond		hardness 10

Various equipment is required for rock collecting, both in the field and at home. The rock collector will need a rucksack, hammers, chisels, magnifying glass, penknife, newspapers, boxes, and soft paper or cotton wool for delicate specimens. A vial of acid with a dropper is useful. Gloves are necessary, as are goggles, and a helmet for collecting under high cliffs. A compass, altimeter, map, notebook, labels, pencil and markers are necessary. It is important to record as much information as possible about the find in the field.

Once the specimens have been brought home, facilities for dealing with them are needed. They must be cleaned using water and brushes (soft or hard as required). A punch, small spatulas and pincers must be on hand, along with a magnet, a mortar and pestle or a piece of porce-

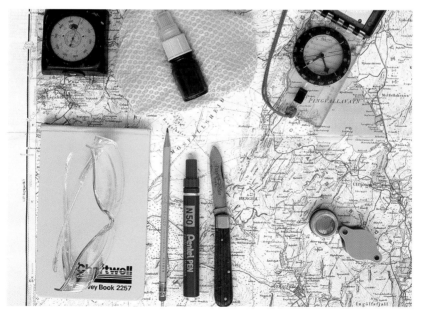

Equipment indispensable in field work.

Boxes etc. for handling and storing specimens.

lain, dilute hydrochloric acid, and rubber gloves for handling the acid. A binocular is very useful, but rather expensive. Small boxes are the best way of storing rock specimens. Various household containers may be re-used for this purpose.

It is important to keep the collection in good order, so that all specimens that are to be kept are clearly marked. Sometimes the identification may be doubtful. A better identification may be achieved later, but this is of no use if the location of the find has been forgotten.

Tools for collecting, cleaning and testing specimens.

Rock-forming minerals

The minerals which crystallise from magma are known as rock-forming or primary minerals, as distinct from those which are formed by alteration, or are deposited by geothermal water or gaseous emanations of solidifying magma. The rock-forming minerals in igneous rock are mainly silicates, i.e. they contain silica (SiO_2). In addition to quartz, these include feldspar, pyroxene and olivine. Mica (biotite) and amphibole (hornblende), which are silicates whose crystals also contain hydroxide, are common rock-forming minerals in igneous rock, but uncommon in Iceland.

Quartz occurs both as a rock-forming mineral and as an amygdule. This is discussed in the section on quartz minerals. Feldspar, pyroxene and amphibole also occur as alteration minerals, which form under high-temperature conditions. Some of these have special names, for instance the variants of pyroxene (hedenbergite) and amphibole (actinolite) which are discussed in the section on high-temperature minerals. Magnetite, also a primary mineral in igneous rock, is discussed with ore minerals.

Some rock-forming minerals occur as large crystals in otherwise cryptocrystalline or microcrystalline volcanic rock. These large crystals are called phenocrysts, and rock containing phenocrysts is called porphyritic. Weathering will sometimes free phenocrysts from the rock, and one may find individual perfect crystals.

Finally, in microcrystalline basalt vesicles may be found where the primary minerals of the rock have grown in gas bubbles or pipes, so that the crystals are bigger than in the matrix. This is called basalt-pegmatite.

Plagioclase phenocryst in basalt. From *Reyðarfjörður, East Iceland*. Size: *largest crystal 1.8 cm.*

FELDSPAR

Crystal system: monoclinic, triclinic
Hardness: 6–6½
Specific gravity: 2.61–2.76

Cleavage: perfect
$NaAlSi_3O_8$ - $CaAl_2Si_2O_8$ and $KAlSi_3O_8$

DESCRIPTION: Feldspars are a group of about 20 minerals of similar properties. Feldspar minerals are aluminium silicates which are generally combined with potassium (K), sodium (Na) or calcium (Ca). Feldspars are divided into two main subgroups, potassium feldspars (sanidine and orthoclase) and plagioclase (albite-anorthite compositional series). Crystals are flat or columnar and granular, with parallel cleavage. Twinning is common. Colour is normally white or greyish-white, while some types are pale yellow or pink. The streak is white. Feldspar is opaque or translucent with conchoidal or uneven fracture and vitreous or pearly lustre.

OCCURRENCE: Feldspar forms more than half the earth's crust. It is the commonest primary mineral in all igneous rock. Calcium-rich plagioclase is the most abundant feldspar in Iceland. Feldspar phenocrysts are common in basalt. They are found either singly or in clusters,

and they may be over 1 cm across. In some porphyritic basalt, feldspar phenocrysts comprise 30–50% of the rock mass. An example of this is Hrímalda north of the Vatnajökull glacier, where the weathered surface is littered with loose phenocrysts. Rhyolite with feldspar phenocrysts is also common, e.g. around the Landmannalaugar geothermal area in the southern uplands.

NAME: The name *feldspar* is from the German *Feldspat* (*feld* = region, *spat* = crystal with perfect cleavage), indicating the fact that this is a very common mineral.

VARIANTS: *Albite* is a sodium-rich plagioclase. It is a primary mineral in acid rocks, and can also form during alteration of plagioclase in high-temperature geothermal systems, where the temperature exceeds 200°C. Potassium feldspar (adularia) also forms in high-temperature systems, but tends to form in cavities and fissures. It is white, with rhomboid prismatic surfaces.

Plagioclase crystals weathered from porphyritic hyaloclastite. From *Ölfus, South Iceland*. Size: *c. 5 mm.*

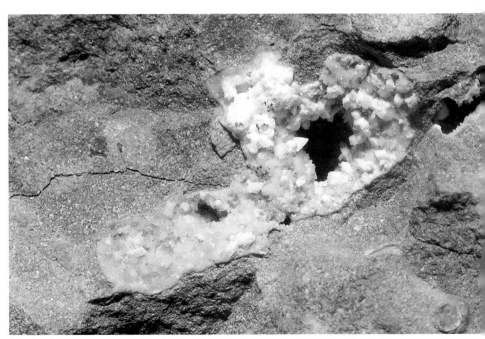

Potassium feldspar as an amygdule in rock, altered at high temperature. From *West Iceland*. Size: *1–2 mm.*

From *Southeast Iceland*. Size: *15 x 9 cm.*

PYROXENE

Crystal system: monoclinic
Hardness 5½–6
Specific gravity: 3.4

Cleavage: perfect
$(Ca,Mg,Al,Ti)_2(Si,Al)_2O_6$

DESCRIPTION: Pyroxene is a compositional series of minerals. Augite is the commonest form of pyroxene in Iceland. It is black or dark green in colour, and forms prismatic crystals with vitreous lustre and perfect cleavage. It is translucent or transparent, streak is grey or green, and fracture uneven.

OCCURRENCE: Pyroxene is one of the principal rock-forming minerals in basalt and gabbro. It is found as phenocrysts in some varieties of basalt and in ankaramite, and may be as much as 0.5 to 1 cm across. Exceptionally larger phenocrysts are found in gabbro.

NAME: *Pyroxene* is from the Ancient Greek (*pyros* = fire, *xenos* = unknown). The second element of the name is derived from the ancient belief that lava absorbed pyroxene by chance. *Augite* is from the ancient Greek (*auge* = bright ray), in reference to the shininess of fresh fractures in the mineral.

Pyroxene phenocrysts in hyaloclastite. From *Mýrdalur, South Iceland*. Size: *largest crystal 1 x 1 cm.*

Pyroxene crystals weathered from hyaloclastite. From *Ölfus, South Iceland*. Size: *largest crystals 5 mm.*

From *Miðfell near Þingvallavatn, Southwest Iceland.* Size: *largest crystals 7 mm.*

OLIVINE

Crystal system: orthorhombic
Hardness: 6½ –7
Specific gravity: 3.2–4.3

Cleavage: indistinct
$(Mg,Fe)_2SiO_4$

DESCRIPTION: Olivine is a compositional series of minerals based on the proportion of magnesium to iron. It forms yellowish-green to dark green granular or prismatic crystals, generally small and difficult to identify with the naked eye, but several millimetres across when it forms phenocrysts. Olivine is transparent or translucent with vitreous lustre and streak is white. Fracture is conchoidal, and cleavage indistinct.

OCCURRENCE: Magnesium-rich olivine is a common rock-forming mineral in basalt, but generally only in small quantities. It is normally the first mineral to crystallise when magma solidifies. Picrite is a type of rock with large amounts of olivine phenocrysts. Olivine has a high specific gravity, and where it weathers from rock it may remain to form olivine-rich sand, for instance at Búðir on the Snæfellsnes peninsula. Olivine also occurs in rhyolite, in which case it is rich in iron.

NAME: *Olivine*'s name is derived from the characteristic olive-green hue of the mineral.

VARIANTS: *Iddingsite* is a form of olivine formed by alteration or weathering. Reddish or even black, it is common in slightly altered basalt.

Olivine crystals weathered from hyaloclastite. From *Ölfus, South Iceland*. Size: *largest crystals 6 mm.*

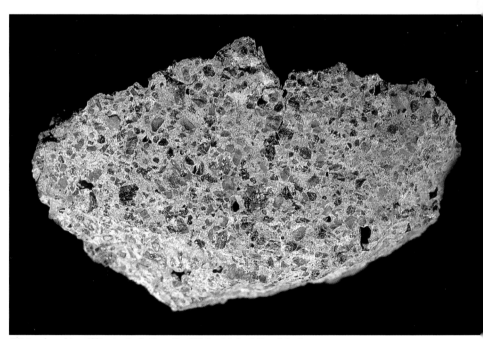

Olivine altered into iddingsite (red). From *Eyjafjöll, South Iceland*. Size: *8.3 x 5 cm.*

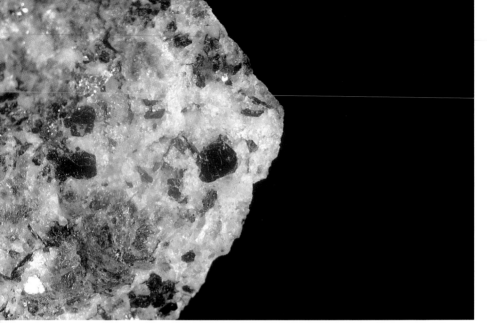

From *Southeast Iceland*. Size: *largest flake 2 mm.*

MICA (BIOTITE)

Crystal system: monoclinic
Hardness: 2½–3
Specific gravity: 2.7–3.3

Cleavage: perfect
$K(Mg,Fe)_3(Al,Fe)Si_3O_{10}(OH,F)_2$

DESCRIPTION: Mica is a group of minerals of which biotite is one member. It forms thin flat crystals or irregular flakes. Colour of biotite is very dark brown or even black and streak is white. Mica has pearly lustre, and the flakes are brittle. The flake form, due to perfect cleavage, is the principal identifying feature, along with the colour.

OCCURRENCE: In Iceland biotite is rare and forms small crystals. Minor amounts have been found in intrusions where it has precipitated from magmatic liquids at a late stage of solidification.

NAME: *Mica* is from the Latin (*mica* = crumb). *Biotite* derives from the name of French physicist Jean Biot (1774–1862).

From *Breiðdalur, East Iceland.* Size: *length of crystals 1–3 mm.*

AMPHIBOLE (HORNBLENDE)

Crystal system: monoclinic
Hardness: 5–6
Specific gravity: 3.2

Cleavage: perfect
$Ca_2(Mg,Fe)_4Al(Si_7Al)O_{22}(OH,F)$

DESCRIPTION: Hornblende is the commonest mineral of the amphibole group of silicates. It is black with vitreous lustre, and opaque. The crystals are hexagonal with end surfaces forming an obtuse angle. Cleavage is perfect, and the cleavage surfaces are shiny. Streak is white or grey. In Iceland hornblende forms single crystals, generally phenocrysts in andesite or dacite. These are usually very small, only a few millimetres in length.

OCCURRENCE: Hornblende is rare in Iceland, but has been found as phenocrysts in andesite and dacite intrusions in Breiðdalur in the east and Króksfjörður in the west. Actinolite, which also belongs to the amphibole group, is a common alteration mineral in high-temperature geothermal systems.

NAME: *Hornblende* is from the German: *horn* = hard, while the *-blende* ending is a reference to translucency, as some forms of hornblende are translucent.

ROCKS

In Iceland only two of the three major groups of rock are found, igneous and sedimentary. Metamorphic rocks do not occur in Iceland, i.e. those metamorphic rocks which are formed under high pressure and temperature. However, rock exists in Iceland which has been formed at high temperature and low pressure at the margins of large intrusions. Over 90% of Iceland's bedrock is igneous rock, formed by solidification of magma. Sedimentary rocks form when igneous rock undergoes weathering and erosion, and is transported by erosive agencies – wind, water, glaciers and sea. During transport, the material is sorted and deposited under various conditions to form new rock under pressure and by chemical processes.

The bulk of igneous rock in Iceland is extrusive, formed on the surface of the earth. Rock of the same composition may solidify deep in the earth, forming intrusions where the magma penetrates into older rocks and solidifies. Igneous rock comprises relatively few minerals, such as pyroxene, olivine and magnetite, which are dark in colour, and feldspar and quartz which are light or translucent. The colour of rocks depends upon the proportion of light- and dark-coloured minerals. Hence basalt is dark, while rhyolite is light.

Magma sometimes solidifies so rapidly that little crystallization takes place, while the remainder forms glass of the same chemical composition as the magma. This occurs due to air-cooling on the surface of lava flows, and due to cooling by cold rock walls where magma fills cavities, but the bulk of glassy rock forms where magma is extruded under water, the sea or a glacier. Iceland's hyaloclastite formations are of this kind.

In rock which has solidified on the surface, e.g. ordinary basalt lava, the minerals are generally small in size, and in some cases they are difficult to identify without a magnifying glass or microscope. In plutonic rock, such as gabbro and the like, minerals may be fairly large and easily identifiable with the naked eye.

The diagram below shows the relationship between the main types of igneous rock, and the differences between them. The size of the crystals is largely due to different rates of cooling, but also due to varying composition. Some rock is more-or-less pure glass, without crystals. In cryptocrystalline rock, normally quite glassy, crystals are hardly discernible unless the sample is magnified under a microscope. Other rock types are microcrystalline, finely crystalline or coarsely crystalline, according to the size of the crystals and diminishing glass content. Coarsely crystalline rock is completely crystallized.

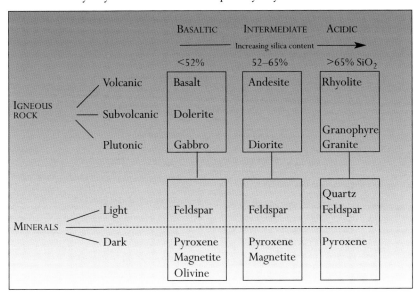

When identifying igneous rock, the minerals must first be identified, and the proportions assessed. This is clearly difficult in the case of crypto- or microcrystalline rocks. Colour and cleavage are important clues. Basalt and andesite are grey or dark in colour, while volcanic rocks with a higher silica content are generally light in colour and characteristically flow-banded and cleave easily. It is often necessary to use a microscope to identify igneous rock. Then a thin section is made, and minerals and their proportions can be identified, even in cryptocrystalline rock.

From *Southeast Iceland*. Size: *8.5 x 5.2 cm.*

GABBRO

DESCRIPTION: Gabbro is a coarsely-crystalline plutonic rock, of the same composition as basalt. Its principal rock-forming minerals are plagioclase-feldspar, pyroxene, magnetite and sometimes olivine. These minerals are generally easily identifiable in hand specimens, except magnetite, which is the smallest of them. The minerals are dark in colour with the exception of feldspar, and so the rock is normally dark or greenish, especially if it has undergone alteration. Gabbro may display banding, if light and dark coloured minerals have separated during crystallization of the magma.

ORIGIN AND OCCURRENCE: Gabbro has solidified in the crust as intrusions. It is commonest in the southeast, e.g. at Eystrahorn and Vestrahorn, and on the Snæfellsnes peninsula in the west, e.g. at Þorgeirsfellshyrna and Kolgrafamúli. Layered gabbro may be seen at Músarnes at the end of the Kjalarnes peninsula in the southwest.

NAME: *Gabbro* is an Italian place-name.

VARIANTS: *Anorthosite* is pale grey, formed almost exclusively of plagioclase-feldspar, of the same composition as in gabbro. In Iceland anorthosite is found in large expanses on Hrappsey island, Breiðafjörður, in the west. The name *anorthosite* is derived from *anorthite*, i.e. calcium-rich plagioclase, which is its principal mineral.

Gabbro from a volcanic bomb. From *Tjarnarhnúkur near Hengill, Southwest Iceland*. Size: *2.5 x 3 cm.*

Anorthosite. From *Hrappsey in Breiðafjörður, West Iceland*. Size: *5.7 x 8.1 cm.*

From *Skagaströnd, North Iceland*. Size: *8.5 x 7.8 cm.*

DOLERITE

DESCRIPTION: Dolerite is a dike rock of the same composition as basalt. In coarseness it is between basalt and gabbro, composed of the same rock-forming minerals. It is grey or dark in colour, and finely crystalline, although the principal minerals may be distinguished with a magnifying glass, or even with the naked eye. When altered it becomes darkish or greenish, and the mineral constituents are less distinct. This rock is sometimes called diabase (basalt which cuts through i.e. dike basalt).

ORIGIN AND OCCURRENCE: Dolerite occurs mainly in thick dikes, e.g. on Viðey island off Reykjavík, or in smaller intrusions, for instance in Esja and Stardalshnúkur in Mosfellssveit, both adjacent to Reykjavík.

NAME: *Dolerite* is from the Ancient Greek *doleros* (= deceptive), in reference to the dark variant of the rock in which the minerals are scarcely distinguishable.

From *Lýsuskarð, Snæfellsnes Peninsula, West Iceland*. Size: *5.3 x 7 cm.*

DIORITE

DESCRIPTION: Diorite is a coarsely-crystalline, intermediate plutonic rock equivalent to andesite and dacite in composition. Its rock-forming minerals are plagioclase-feldspar, pyroxene, hornblende and quartz. In Icelandic diorite, the dark mineral is mainly pyroxene, while in other diorite it is normally hornblende, a mineral similar to pyroxene, but containing hydroxide.

ORIGIN AND OCCURRENCE: Diorite is rare in Iceland. In Lýsuskarð above Lýsuhóll on the Snæfellsnes peninsula an intrusion of intermediate rock occurs which may be classified as diorite. A dike similar to diorite in texture and composition, containing hornblende, is found at Króksfjörður in the west.

NAME: *Diorite* is from the Ancient Greek *diorizein* (= separate), in reference to the fact that the minerals are sufficiently coarse in texture to be distinguishable.

Microgranite. From *Hornafjörður, Southeast Iceland*. Size: *8 x 5 cm.*

GRANOPHYRE–MICROGRANITE

DESCRIPTION: Granophyre is a finely or coarsely crystalline acid plutonic rock, equivalent to rhyolite in composition. It is normally light or greyish in colour, as its principal minerals are feldspar and quartz. Dark minerals, magnetite, pyroxene and olivine, are minor constituents, but these do not account for much more than 5% of the rock mass. Microscope examination reveals fine-scale intergrowth of quartz and feldspar, which is an identifying feature of granophyre. If this feature does not occur, the rock is called microgranite. Granophyre is related to granite, a common rock in the continental crust. It is far more coarse-grained than granophyre.

ORIGIN AND OCCURRENCE: Granophyre has solidified in large intrusions. It occurs in the same areas as gabbro, e.g. in Eystrahorn and Vestrahorn in the southeast, and in Lýsuhyrna on the Snæfellsnes peninsula. Microgranite and granite do not occur in Iceland, so far as is known, except at one site, i.e. in the inner reaches of the Slaufrudalur valley in Lón in the southeast.

NAME: *Granophyre* derives from the Ancient Greek *grano* (= granular) and *phyr* (= phyric, i.e. containing phenocrysts). However, phenocrysts would no longer be regarded as a defining characteristic of the rock.

46

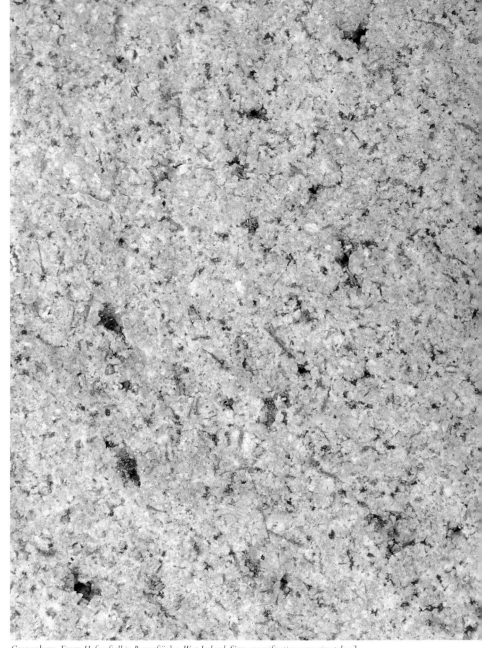

Granophyre. From *Hafnarfjall in Borgarfjörður, West Iceland*. Size: *magnification approximately x2*.

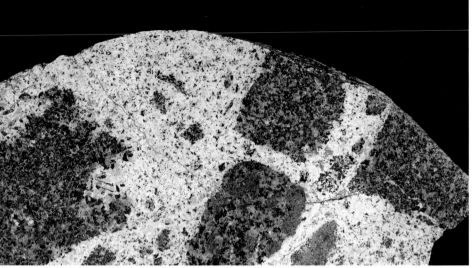

From *Eystrahorn, Southeast Iceland.* Size: *width 6.5 cm.*

GABBRO/GRANOPHYRE COMPOSITE ROCK

DESCRIPTION: Gabbro and granophyre are sometimes found intermingled as a composite rock. The gabbro forms rounded or corroded inclusions in the granophyre. The smallest may be less than 1 cm across, the largest several metres. The rims of larger gabbro inclusions and the smallest ones often become finely crystalline (dolerite) and pale in colour, i.e. the colour changes from dark to pale green.

ORIGIN AND OCCURRENCE: Most of the large intrusions have been formed by magma batches being intruded succes-sively, sometimes before the first were fully crystallized. If the successive intrusions are of different composition, a composite rock can be formed. Basic magma, which is hotter than acidic magma, will become finely crystalline at the contact due to the cooling effect. It can also mingle with the acidic magma to form a mixed melt that solidifies as diorite. The best-known location of gabbro/granophyre composite rock in Iceland is at Eystrahorn in the southeast.

BASALT

DESCRIPTION: Basalt is a glassy, micro-crystalline or finely crystalline basic rock, whose silica (SiO_2) content is under 52%. Basalt varies greatly, and its different varieties have specific names. In scoria and hyaloclastite, glassy basalt predominates. Fresh, finely crystalline basalt is known in Iceland as *grágrýti* (greystone), while microcrystalline, dense basalt is called *blágrýti* (bluestone).

Groundmass: The principal minerals in basalt are plagioclase-feldspar, which comprises 40–50% of the rock, pyroxene, another 40–50%, and metallic minerals, i.e. oxides of iron and titanium. In addition to this, basalt may also contain some quantity of olivine. These minerals form the groundmass of the rock; they are rarely distinguishable in hand specimens due to their small size.

Phenocrysts. Phenocrysts of plagioclase-feldspar, olivine and pyroxene are common in basalt – especially the first two. Plagioclase is white, pyroxene black, and olivine green or greenish-yellow. Phenocrysts tend to sink gravitively in molten basalt, both in lava flows and in pillow lava. This applies especially to the dark phenocrysts, but also feldspar phenocrysts, which are then rich in calcium, with a higher specific gravity than the melt.

Some varieties of basalt have specific names, such as picrite, olivine basalt, tholeiite and ankaramite. Basalt with abundant feldspar phenocrysts is known as porphyritic basalt. Varieties of basalt are distinguished by the composition of the primary minerals and their proportions or, more commonly in the present day, by the whole rock composition. This is done by examining a thin section under a microscope, or by chemical analysis. In the field, rock types may often be identified by their external features and texture. External features include e.g. type of weathering, whether lava layers are simple or compound, and what secondary minerals are present. With regard to structure, coarseness, type of phenocrysts and flow-banding are identifying characteristics.

ORIGIN AND OCCURRENCE: Basalt is an extrusive or hypabyssal rock, which is by far the commonest rock in Iceland. Most lava fields in volcanic zones, and lava layers in older formations, as well as hyaloclastite, are of basaltic composition. Most dikes are also basalt, except for the thickest ones.

NAME: *Basalt* is said to derive from the Ethiopian language, in which it refers to a black iron-rich rock. The name was first used in writing by the Elder Pliny, a Roman historian of the first century AD.

Fresh olivine basalt (*grágrýti*). From *Geldinganes, Reykjavík, Southwest Iceland. Size: 4 x 3 cm.*

OLIVINE BASALT

DESCRIPTION: Olivine basalt is micro-crystalline or finely crystalline, grey when fresh (Icelandic *grágrýti*), but turns very dark when altered, and weathers easily. The surface then becomes crumbly and uneven. Spheroidal weathering is common, and sometimes causes the rock to break up into concave flakes. Olivine is often visible to the naked eye. There is little or no flow-banding. Vesicles often form pipes extending upwards from the base of olivine basalt lava flows. The flow tops are vesicular, and the surface may be ropy. In trains of vesicles, the rock may be coarsely crystalline, and the minerals may be distinguishable with the naked eye. Rock of this kind is called basalt-pegmatite. Altered olivine basalt contains various zeolites, but quartz is absent.

ORIGIN AND OCCURRENCE: Olivine basalt is a common extrusive rock, forming both compound and simple lava flows. Compound structure is most distinct in shield lavas. The *grágrýti* around Reykjavík is olivine basalt. The walls of Ásbyrgi in the north and Almannagjá at Þingvellir in the southwest consist of innumerable flow units of olivine basalt. Ancient, somewhat altered olivine basalt forms the mountains on both sides of Ljósavatnsskarð in the north, and clearly identifiable lava groups characteristic of compound lavas may be seen on Múlafjall in Hvalfjörður in the west. Olivine basalt is believed to form by partial melting of mantle rock, or by fractional crystallization of picrite in the earth's crust.

50

Basalt-pegmatite. Large crystals of pyroxene in olivine basalt. From *Geldinganes near Reykjavík*. Size: *4 x 3 cm.*

Altered olivine basalt. From *Breiðdalur, East Iceland*. Size: *13 x 8 cm.*

From *Hvalfjörður, West Iceland.* Size: *13.5 x 24 cm.*

THOLEIITE

DESCRIPTION: Tholeiite (Icelandic *blá-grýti*) is microcrystalline and dark in colour, sometimes grey when fresh. The surface of weathered tholeiite is grey or greyish-brown, disintegrating into rather sharp-edged rock pieces or blocks. Tholeiite is generally flow-banded, especially in the middle of lava flows, sometimes causing it to split into slabs. The flow-banding appears as brownish, flat streaks when viewed in vertical section. In the flow bands, the rock is more vesicular and the crystals larger than in the microcrystalline groundmass. Tholeiite lavas are generally fairly vesicular, especially at the top, and they usually have proportionately thick surface rubble. Secondary minerals only occur at considerable depth in the lava sequence of the basalt formation, and these are predominantly silica minerals or silica-rich zeolites.

ORIGIN AND OCCURRENCE: Tholeiite constitutes about 50% of the basalt formation in Iceland. It forms either groups of thin lava flows around central volcanoes, attributable to their frequent fissure eruptions, or individual thick lava flows erupting from fissures in the swarms that pass through them. Tholeiite is formed by the fractional crystallization of olivine basalt, i.e. chromite and olivine are subtracted and precipitated in magma chambers in the earth's crust.

NAME: *Tholeiite* (coined in 1840) derives from *Tholei*, the name of a rock on the Rhine.

From *Reyðarfjörður, East Iceland*. Size: *16 x 12 cm*.

PORPHYRITIC BASALT

DESCRIPTION: Most varieties of basalt may contain phenocrysts, i.e. large crystals in a microcrystalline or finely crystalline groundmass. If they make up less than 5% of the rock, it is termed sparsely-porphyritic. If the phenocrysts are more abundant, the rock is called porphyritic. The phenocrysts are most commonly feldspar, and this is what is generally meant by porphyritic basalt. Dark-coloured olivine and pyroxene phenocrysts also occur. If the dark minerals predominate among the phenocrysts, the rock is termed picrite or ankaramite. Phenocrysts are often composed of many individual crystals, including some dark minerals. In this case, the rock is called glomero-porphyritic. Some porphyritic basalt contains considerable quantities of large feldspar phenocrysts (0.5 cm and over),

comprising up to 50% of the rock mass. This is referred to as cumulate. Such basalt is very easily identified. It forms thick flows, and fractures into large blocks. It is sought after for use in breakwaters and harbour walls, where currents are strong.

ORIGIN AND OCCURRENCE: Porphyritic basalt comprises about 10% of the basalt formation in Iceland. It often forms groups of lavas which can be traced over large areas, and are useful marker horizons in geological mapping. The phenocrysts grow in subsurface magma chambers, and are carried to the surface during eruptions. They sometimes bear little relation to the rock in which they are found. In this case they are called xenocrysts which means "foreign crystals".

53

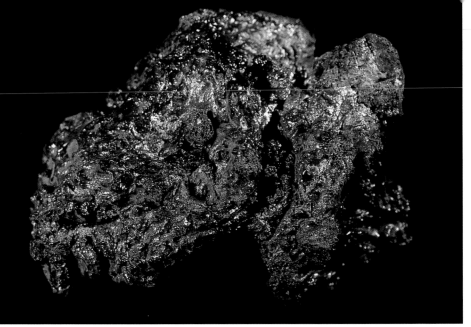

Black scoria with glassy surface. From *Reykjanes, Southwest Iceland.* Size: *11 x 8 cm.*

SCORIA – LAPILLI – ASH – PELE'S HAIR

DESCRIPTION: Scoria is glassy basalt or andesite, highly vesicular or frothy. It is either black or reddish, especially where the scoria pile is thick. A multi-coloured glassy surface is often seen on black scoria. Compressed lumps of scoria and lava are known as spatter. Volcanic material in the form of grains and fine dust is called lapilli (Latin *lapillus* = small stone) and ash. Delicate, transparent glassy needles, which are deposited around craters, are known as Pele's hair. These shiny needles often have a yellowish sheen. The name Pele's hair is Hawaiian. In the religion of the indigenous people, Pele was the goddess of fire.

ORIGIN AND OCCURRENCE: Scoria forms in volcanic craters from fragments of glowing lava which fall to the ground around the vent. It piles up quickly, and remains hot for a long time. Hot, humid air that ventilates through them leads to oxidation of ferrous minerals, giving a reddish colour. Basaltic lapilli and ash form in explosive eruptions where water has entered the vent and expanded as it forms steam. The Icelandic volcanoes which have produced the largest amounts of basalt ash are Katla in Mýrdalsjökull and Grímsvötn, under the Vatnajökull glacier.

Red scoria. From *Reykjanes, Southwest Iceland*. Size: *11 x 8 cm*.

Black lapilli. From *Sólheimasandur, South Iceland*.
Size: *largest particle 1.6 x 1.3 cm*.

Pele's hair. From *Krafla, Northeast Iceland*. Size: *6.5 x 4 cm*.

55

Palagonite tuff is generally stratified. From *Reykjanes, Southwest Iceland*. Size: *27 x 18 cm.*

HYALOCLASTITE AND PILLOW LAVA

DESCRIPTION: Hyaloclastite (Icelandic *móberg*) and pillow lava are glassy variants of basalt formed in submarine or subglacial eruptions. Hyaloclastite forms when water flows into a crater, causing steam explosions and shattering of the molten lava. The magma does not crystallize, but solidifies as glass and breaks down into shards. When basaltic glass is altered, it turns brown. This is known as palagonite. Pillow lava forms in deep water, or subglacially. The rim of the pillows, which cools most rapidly, has a black glassy surface known as tachylite. In pillow lavas the pillows are generally densely packed, forming piles in the lower part of hyaloclastite mountains. Pillow lava may also form ridges or thick, flat expanses.

ORIGIN AND OCCURRENCE: Hyaloclastite and pillow lava are common in volcanic formations from the Ice Age. The two different rock types largely depend on the depth of water in which they formed. Hyaloclastite tuff is practically pure glassy particles, which have become consolidated into rock. Some has piled up above water around the vent (Surtsey island), but most has been deposited under water on the slopes of hyaloclastite mountains, or adjacent to them. Layers of such palagonite tuff may be seen e.g. on Hengill near Reykjavík. Hyaloclastite containing abundant lithic fragments is called breccia. Mosfell, a prominent hill in Mosfellssveit south of Esja, is entirely composed of pillow lava. Hyaloclastite mountains are a characteristic feature of the landscape in Iceland's volcanic zones, as all the steep-sided mountains were formed subglacially. Where the eruption breaks though the glacier, lava layers can form on top of the hyaloclastite, forming a tuya or table mountain (Icelandic *stapi*), e.g. Herðubreið in the northeast.

Pillow lava. From *Kálfstindar, South Iceland*. Size: *diameter of pillows about 1 m.*

Palagonite. From *Brúardalir, northeast of Vatnajökull*. Size: *10 x 3 cm.*

Breccia. From *Krýsuvík, Southwest Iceland*. Size: *21 x 16 cm*.

Altered hyaloclastite. The glassy fragments of the groundmass have been largely altered into zeolites. From *Húsafell, West Iceland*. Size: *4.5 x 5 cm*.

Tachylite. From the *Öræfi district, Southeast Iceland*. Size: *6.5 x 7 cm*.

From *Miðfell near Þingvallavatn, Southwest Iceland. Size: 15 x 13 cm.*

PICRITE

DESCRIPTION: Picrite is a basalt with a proportionately low silica content, and large quantities of green olivine phenocrysts. The groundmass is finely crystalline, grey or pale grey. The rock is often quite vesicular. The groundmass is, as in basalt, mainly plagioclase-feldspar, pyroxene, olivine and ore minerals. Phenocrysts of green or yellowish-green olivine, usually under 0.5 cm in diameter, comprise more than 20% of the rock. Also characteristic are small (discernible by microscope), sparsely distributed phenocrysts of black or dark rusty-brown chromite $((Mg,Fe)Cr_2O_4)$. The rock also generally contains a considerable amount of feldspar and pyroxene phenocrysts.

ORIGIN AND OCCURRENCE: Picrite is an extrusive rock, fairly rare in Iceland. It is a member of the tholeiite rock suite. Picrite lava flows from the early Holocene occur on the Reykjanes peninsula in the southwest. The best examples are Háleyjabunga and Lágafell. The southwestern part of Miðfell, at the east of lake Þingvallavatn, is an easily accessible example. Picrite is the most primitive variant of basalt, i.e. most similar to the undifferentiated magma in the mantle.

NAME: *Picrite* derives from picotite (from *pico* = tiny), a mineral related to chromite, which occurs in all picrites.

Pyroxene (black) and olivine (green) phenocrysts in ankaramite. From *Eyjafjöll, South Iceland*. Size: *4.5 x 6 cm.*

ANKARAMITE

DESCRIPTION: Ankaramite is a highly porphyritic alkalic basalt containing large amounts of dark minerals. The groundmass is finely crystalline or microcrystalline. The rock is greyish, either vesicular or dense. The phenocrysts in ankaramite are pyroxene and olivine, as well as some feldspar. Pyroxene occurs in larger quantities than olivine.

ORIGIN AND OCCURRENCE: Ankaramite is found in Iceland in the same regions as alkali basalt, especially around Eyjafjallajökull in the south. It is found both as lava and as dikes. As dike rock it is best known at Hvammsmúli south of Eyjafjallajökull. Ankaramite is believed to form by partial melting of perdotite in the mantle beneath the non-rifting flank volcanic zones.

NAME: *Ankaramite* derives from the place-name *Ankaramy* in Madagascar.

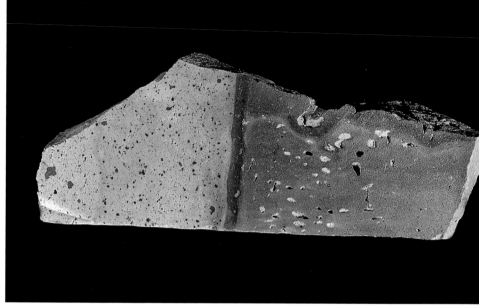

From *Breiðdalsvík, East Iceland*. Size: *length 20 cm*.

BASALT/RHYOLITE COMPOSITE ROCK

DESCRIPTION: Basalt and rhyolite are often found in association, especially in ignimbrite, but also in lavas and dikes. In ignimbrite, basalt forms vesicular lapilli or cinder-like particles. In lava it forms small inclusions or streaks, and sometimes a coherent layer at the bottom of acid lava flows. In dikes, basalt is found at the margins, and the rhyolite in the middle. At the transition there is a zone, 0.5–1 m across, with basalt forming drops and veins in the rhyolite. Various shades of colour occur in the transition zone, indicating that hybrid magma of andesite-dacite composition has formed.

ORIGIN AND OCCURRENCE: Composite rock of this type forms where basaltic and acidic magma erupt simultaneously. Basaltic magma has intruded into acidic magma at depth, and the two erupted together. Ignimbrite is very often of this type. A famous example is the Skessa ignimbrite in Breiðdalur in the east. The Dómadalshraun and Námshraun lavas near Landmannalaugar are of this type. The best example of a composite dike is at Streitishorn south of Breiðdalsvík in the east.

Andesite. Silica content c. 54%. From *Borgarfjörður, West Iceland*. Size: *13 x 9 cm.*

ANDESITE AND DACITE

DESCRIPTION: Andesite and dacite are intermediate rocks with silica content between 52% and 67%. Both are microcrystalline and dark, or greyish. They invariably are flow-banded and split into slabs or thin cleavage plates when weathered. The flow-banding differs from flow-banding in tholeiite, as it often undulates. As the rock becomes more acid, the curving of the flow-banding becomes more pronounced. Fracture surfaces are irregular with sharp edges, and small offsets where the rock cleaves along flow bands. The rock is often reddish due to oxidation. One of the characteristics of the rock is that it gives a clinking sound if struck with a hammer, or if two pieces of rock are struck together. Andesite or dacite lavas have a thick, rubbly flow top, often reddish in colour. Andesite is generally without phenocrysts, but dacite is sometimes porphyritic, with feldspar phenocrysts. The groundmass of both rocks is very fine-grained or glassy, with minute crystals of feldspar, pyroxene and magnetite. The magnetite content of andesite is relatively high. In dacite, quartz may sometimes be found.

ORIGIN AND OCCURRENCE: Andesite and dacite are found in central volcanoes, where they form thick lava flows which cover a relatively small area. Andesite and dacite lavas which have flowed after the Ice Age are found at Heiðarsporður near Mývatn, and at Hekla. Dacite and andesite are believed to be of hybrid origin, i.e. formed by mixing of basic and acid magmas, as the chemical composition is rather variable within the same volcanic unit.

NAME: *Andesite* is derived from the *Andes* mountains in South America, where it was first described. *Dacite* is from the regional name *Dacia* in Transylvania (Romania).

Andesite. From *Hekla, South Iceland*. Size: *2.6 x 7 cm.*

Dacite. Silica content 64–65%. From *Hraunbunga near Mývatn, North Iceland*. Size: *11 x 5 cm.*

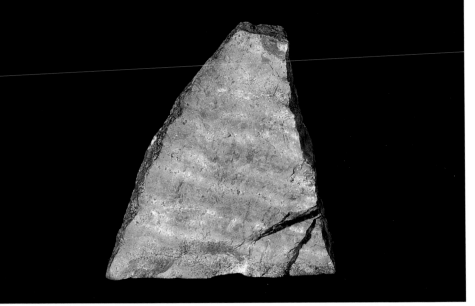

From *West Iceland*. Size: *9 x 10 cm.*

RHYOLITE

DESCRIPTION: Rhyolite is an acid rock containing over 67% silica. It is pale grey, yellowish or pink, and microcrystalline, so individual crystals can only be distinguished under a microscope. Rhyolite occurs as lavas often 50–100 m thick and as pyroclastic flows, but also as dikes and smaller laccoliths. At the surface and margins of such formations, a black glassy obsidian or pitchstone is formed. Rhyolitic magma is viscous, so lavas flow only a short distance from the vent. Rhyolite lavas are flow-banded, sensitive to weathering, and cleave into slabs and plates. A yellowish-brown or brown colour along the cleavage surfaces is due to oxidation of iron. Rhyolite is often porphyritic, especially with phenocrysts of sodium-rich feldspar, but also iron-rich augite and olivine may occur. The groundmass is microcrystalline and glassy, with tiny feldspar and quartz.

Magnetite occurs as minute crystals in rhyolite and pyrite is commonly present. Spherulitic texture is very characteristic of extrusive rhyolite.

ORIGIN AND OCCURRENCE: Rhyolite is a common rock in Iceland. Among recent rhyolitic lavas are e.g. the obsidian flows near Landmannalaugar. Rhyolite occurrences from the Pleistocene (Ice Age) include Ljósufjöll on the Snæfellsnes peninsula, Hágöngur and Kerlingarfjöll in the central highlands, and Móskarðshnúkar at the east of Esja, adjacent to Reykjavík. Rhyolite is believed to form by partial melting of basalt in the lower part of the crust.

NAME: *Rhyolite* is from the Ancient Greek *rhyax* (= lava flow), and has reference to the frequent flow-structure. The term *liparite*, derived from the *Lipari* islands off Sicily, was commonly used by continental petrographers as a synonym for rhyolite.

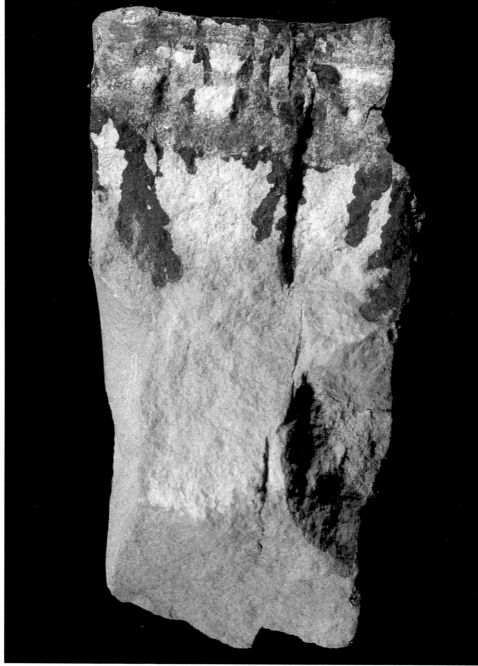

Rhyolite from an intrusion, without flow-banding. Iron oxides have seeped into the rock from fissures, forming dendritic patterns. From *Drápuhlíðarfjall, Snæfellsnes Peninsula, West Iceland.* Size: *8.3 x 4.2 cm.*

From *Hrafntinnusker near Landmannalaugar in the southern highlands*. Size: *9 x 4.5 cm.*

PUMICE

DESCRIPTION: The term pumice refers to vesicular glassy pyroclastics. It is light-coloured and dacitic or rhyolitic in composition. It is frothy, often fibrous, and sufficiently light in weight to float on water until it becomes saturated.

ORIGIN AND OCCURRENCE: Pumice forms by rapid solidification of acidic melt in explosive eruptions. It may be carried by wind far from the eruption site. Several of Iceland's major volcanoes have erupted large quantities of light pumice, e.g. Hekla, Snæfellsjökull, Öræfajökull and Askja. Large expanses of pumice cover the area between Hekla and nearby Búrfell, and thick deposits are found in glacial moraines by the Snæfellsjökull glacier.

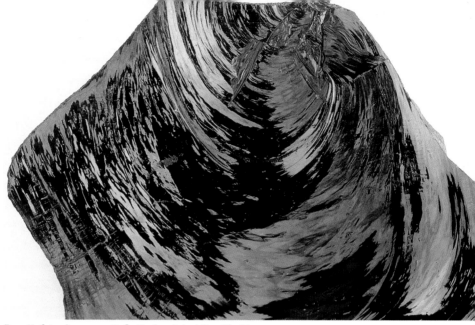

From *Hrafntinnuhryggur near Krafla, Northeast Iceland. Size: 20 x 16 cm.*

OBSIDIAN

DESCRIPTION: Obsidian is the extreme glassy modification of rhyolite. It is black or sometimes dark grey in colour, with a vitreous lustre. It may be streaky due to different shades of colour. When struck, it breaks into sharp-edged pieces with conchoidal fracture. It very often contains feldspar phenocrysts, but is only prized if free of them.

ORIGIN AND OCCURRENCE: Obsidian forms by rapid cooling of rhyolite, e.g. at the edges of dikes or on the surface of lava flows. The best-known occurrences are obsidian lavas in the Landmannalaugar area, of which all but one are porphyritic, and Hrafntinnuhryggur (= Obsidian Ridge) near Krafla.

From *Skarðsheiði, West Iceland*. Size: *width 4.5 cm*.

PITCHSTONE

DESCRIPTION: Pitchstone is a glassy variety of rhyolite. It has a dull or greasy lustre and is generally black, resembling coal, but is sometimes greenish or brownish. It is often brittle and fissured. Pitchstone is altered obsidian, which has absorbed water and lost its vitreous lustre.

ORIGIN AND OCCURRENCE: Pitchstone is associated with rhyolite rocks, invariably forming the marginal layer of acid lava flows and dikes. Well-exposed rhyolite dikes with black pitchstone margins are seen at Hraunfossar in Borgarfjörður in the west, and in nearby Deildargil.

VARIANTS: *Acid hyaloclastite*. When acid magma erupts below a glacier, the melt disintegrates into lava pods and fragments of glassy shards. The erupted material builds up into piles or even mountains. The innermost part of the pod is of lithic rhyolite, with an outer layer of pitchstone, which may form regular columns. The rock in these formations is more glassy and lighter in colour than ordinary hyaloclastite, and the lithic fragments in the breccia are of pitchstone. Freshly formed, the acid hyaloclastite is unconsolidated and perlitic in texture. It becomes indurated upon alteration and absorption of water, to resemble hyaloclastite.

Porphyritic pitchstone. Some 50% of the rock mass consists of feldspar phenocrysts. From *Jökuldalir near Landmannalaugar in the southern highlands.* Size: *4.8 x 7.4 cm.*

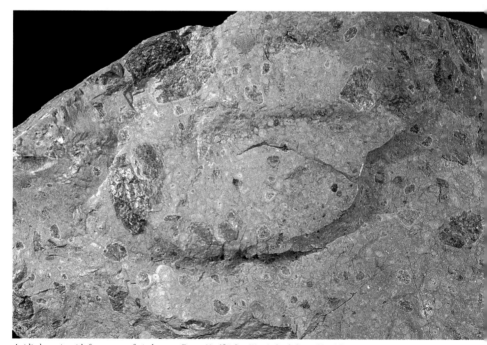

Acidic breccia with fragments of pitchstone. From *Hvalfjörður, West Iceland.* Size: *11 x 15 cm.*

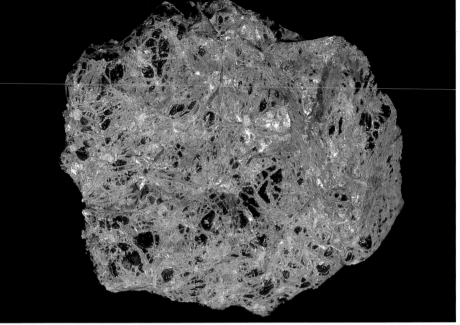

From *Prestahnúkur, West Iceland.* Size: 7 x 4.5 cm.

PERLITE

DESCRIPTION: Perlite is a glassy variety of rhyolite. It is grey, and often crumbly, with a greasy or vitreous lustre, and it generally contains dark spherical grains with concentric cracks. Weathering breaks it down into rounded pebbles or "pearls", hence the name perlite. Perlite has a high water content, of 5% or more. Due to the water content, perlite expands when heated.

ORIGIN AND OCCURRENCE: Perlite has formed mainly where rhyolitic magma has been extruded under water or a glacier, and solidified without losing its primary water content. Perlite is found in quantity at Prestahnúkur west of Langjökull, and in Loðmundar-fjörður in the east.

Dense perlite. From *Prestahnúkur, West Iceland.* Size: *6.6 x 4.5 cm.*

Crumbled perlite. From *Prestahnúkur, West Iceland.* Size: *approximately true size.*

Volcanic bomb with xenolith. From *Krafla, Northeast Iceland.* Size: *9.2 x 11.5 cm.*

XENOLITHS

DESCRIPTION: Xenoliths are fragments of rock which have become detached from the walls of volcanic vents or magma chambers, and carried to the surface with the magma. Rock fragments of this kind differ in composition from the magma, and this is the origin of the term *xenolith* (= foreign stone). Xenoliths may be found loose in scoria, sometimes forming the core of volcanic bombs, but also in lava flows, where they generally occur as coarsely-crystalline inclusions or nests.

ORIGIN AND OCCURRENCE: Xenoliths are of various kinds. In pyroclastics near explosion craters, fragments of rock from the layers immediately below are common. Examples of this are Hverfell at Mývatn, and Kerling north of the Reykjanes lighthouse. Xenoliths of intrusive rocks also may be brought to the surface, having broken off from the walls of magma chambers. Of these, gabbroic xenoliths are most common. They have been found both in pyroclastics and lava flows. Granophyric xenoliths are generally found in pumice and ignimbrite. Xenoliths are common in Iceland. They occur in lava bombs at Grænavatn in Krýsuvík in the southwest, and as small fragments in the scoria of Seyðishólar in Grímsnes in the south. Well-known sites for gabbroic xenoliths in lava are Hraunsvík east of Grindavík on Reykjanes peninsula, and the lava flow from Tjarnarhnúkur in Grafningur south of Þingvallavatn. Granophyric xenoliths are common around Víti at Krafla. This rock was once known as *krablite* (from the name of *Krafla,* pronounced *Krabbla* in Icelandic).

Gabbro xenoliths in pillow lava. From *Miðfell near Þingvallavatn, Southwest Iceland.*
Size of xenoliths about 5 cm.

Welded ignimbrite. From *Skarðsheiði, West Iceland*. Size: *18 x 7 cm.*

IGNIMBRITE

DESCRIPTION: Ignimbrite is composed of dacite or rhyolite. It forms in highly explosive eruptions, when volcanic material cascades down the ground as ash flows. Ignimbrite is usually grey or pale in colour like rhyolite, but becomes reddish due to oxidation around degassing vents. Upon secondary alteration it turns greenish or pinkish. Ignimbrite can be identified by dark glassy lenses, originally pumice which has collapsed and become flattened. These are often 0.5–1 cm thick, and elongated. The matrix is fine-grained glassy powder, consolidated to a various degree. Some ignimbrite is unconsolidated, but more often it is welded into solid rock. It is generally vesicular, the vesicles being drawn out like the collapsed pumice fragments. Flow-banding may occur in thick layers. Xenoliths are always found in ig-

nimbrite. They have broken from the walls of eruption conduits at varying depth, or in the case of intrusive rock fragments been brought up from the level of the magma chambers.

ORIGIN AND OCCURRENCE: In paroxysmal eruptions, a mixture of gas, rock fragments and incandescent volcanic glass in a semi-molten state is expelled into the air. It falls back around the vent due to its weight, and low viscosity causes it to flow freely along the ground as an "ash flow". When the flow settles, it degasses and becomes compressed under its own weight. During compaction vesicular pumice collapses and flattens out. No major ignimbrite eruptions have taken place in Iceland after the Ice Age. Well-known ignimbrite localities are around Húsafell in the west, in Berufjörður in the southeast and in Þórsmörk in the south.

74

Slightly welded ignimbrite. From *Berufjörður, East Iceland*. Size: *12 x 18 cm.*

From *Hvalfjörður, West Iceland.* Size: *2.3 x 3 cm.*

SPHERULITES

DESCRIPTION: Spherulites are small, round, reddish-brown, grey or green bodies which are often found in rhyolite. They are harder than the surrounding rock matrix, and thus may become free upon weathering. When broken through a spherulite often reveals a radial texture and several concentric layers, sometimes of different colours, with occasional spaces between layers. They are frequently 0.5–2 cm in diameter, but much larger spherulites have been found, the largest 10 cm or more across. They are often grown together. Small spherulites, less than 0.5 cm, are very common in rhyolite.

ORIGIN AND OCCURRENCE: Spherulites form where gas is trapped in rhyolitic lava or ignimbrite. Needles of feldspar and quartz grow radially from a common centre. During crystallization, heat is produced that halts the growth temporarily. This process may be repeated several times, so that several concentric rings are revealed if the spherulite is cut open. Well-known locations are Hvaleyri in Hvalfjörður in the west, and Álftavík in Borgarfjörður in the east. According to folklore, spherulites had various supernatural qualities, e.g. they were reputed to stop a flow of blood.

From *Pálsfjall in Vatnajökull, Southeast Iceland*. Size: *10 x 11 cm*.   From *Álftavíkurtindur, East Iceland*. Size: *18 x 20 cm*.

From *Borgarfjörður, West Iceland*. Size: *9 x 6 cm*.

77

Sedimentary rocks are those which have formed by the consolidation of sediment and rock debris. Erosive agents, i.e. wind, water, glaciers and sea, break down rock, transport it away, sort it by size, and finally deposit it as sediment. There are many types of sediment of different origin and composition. Only some main types of sedimentary rock will be discussed here, including those in which fossils or traces of organisms are preserved. Fossils are mainly found in the fine-grained sedimentary rocks, such as claystone and sandstone. Organic remains are rarely found in conglomerate and glacial till, but bones and shells are also found in unconsolidated gravel formations.

Plant remains are commonly found in sedimentary interbeds between lava layers. The plant- bearing interbeds are generally found associated with major sedimentary series, which usually can be traced over tens of kilometres, along the strike of tilted rock sequences. This applies, for instance, to the layers of lignite in the West Fjords and West Iceland, which comprise six series of such strata, interspersed with sequences of lava, 1,000 m or more thick. Plant remains are also found locally where conditions have been favourable for their preservation, e.g. in tuff layers and lake sediments of limited extent. Identifiable plant remains have made it possible to trace the climatic and vegetation history of Iceland from the oldest formations to the Ice Age.

Invertebrate fossils are primarily found in ancient marine sediments. These include molluscs and crustaceans, i.e. shells of bivalves, gastropods and balanides, not to mention such microscopic groups as foraminifera. Bones of marine vertebrates are also found. Shells are preserved whole or in fragments, or sometimes as a cast, if the original calcium carbonate shell has been dissolved. The oldest Icelandic fossil shells date from the end of the Tertiary Period.

The most abundant fossil-bearing strata are at Tjörnes, Northeast Iceland. They formed in the period from the latest epoch of the Tertiary to the mid Ice Age. The fauna of the Tjörnes layers provides evidence of great environmental and climatic change during the period of formation.

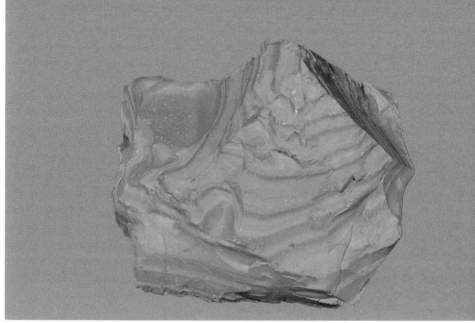

Stratified claystone. From *Hvalfjörður, West Iceland.* Size: *9 x 9 cm.*

CLAYSTONE

DESCRIPTION: Claystone consists of very fine-grained consolidated rock meal, i.e. grain size is considerably less than 0.1 mm. The rock also includes a variable amount of clay minerals. Most of the rock called claystone in Iceland is in fact hardened silt. Claystone usually originates as lacustrine sediment, settling out in sluggish rivers, ponds and lakes, or as marine sediment, carried out to sea by rivers. Claystone is usually distinctly laminated. The layering was originally horizontal. Claystone may contain layers of consolidated tuff and diatomite. It is sometimes varved, i.e. composed of layers of different colour and grain size corresponding to deposits during summer and winter.

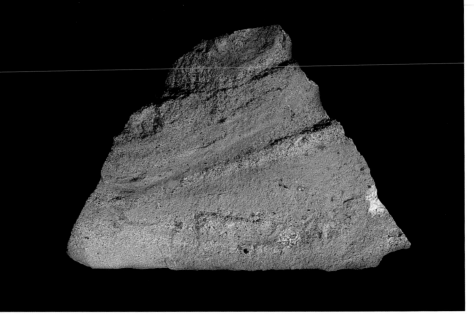

From *Jökuldalur, East Iceland*. Size: *10 x 6.5 cm.*

SANDSTONE

DESCRIPTION: Sandstone consists of consolidated rock debris, with grain size from just under 0.1 mm to 2 mm. It generally consists of particles of the common rock types and their minerals, both rock-forming and alteration minerals. In formations of the Quaternary Period, i.e. since the Ice Age, Icelandic sandstone is generally high in glass content. Sandstone is formed from wind-blown sand, fluvial sediment, and under marine conditions, in the near-shore environment. Due to the basaltic origin Icelandic sandstone is generally dark in colour. It is bedded, consisting of regular layers or more often variously dipping stacks of layers, a feature referred to as cross-bedding. It is produced by changing currents of the transporting agent, water and wind.

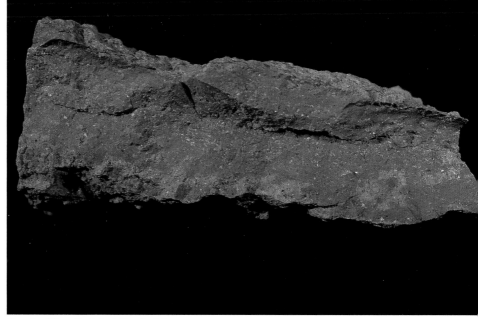

From *Snæfellsnes, West Iceland*. Size: *5 x 2.3 cm.*

INTERBASALTIC BEDS

DESCRIPTION: Sedimentary beds that occur between basalt layers are generally red in colour, of somewhat variable composition and grain size. The interbasaltic beds are ancient soil that accumulated during the interval between the successive lava flows. They consist mainly of glassy tephra, which was transported by wind and reworked by water and subsequently weathered. The time interval between lavas in the basaltic formation averages about 10,000 years. In the warm, damp climate of the Tertiary, this was sufficient to permit weathering of the glassy soil fragments into clay minerals and ferric compounds. Under subsequent burial it was followed by low-temperature alteration of glass and basaltic rock fragments into zeolites and quartz minerals. The red colour is due to iron in an acidic soil, which was oxidized mainly into haematite after being covered by lava. The heat from the lava "baked" the soil underneath and the acidity of the soil is lost over time. Laterite is found in the oldest of Iceland's interbasaltic beds, at the westernmost part of the West Fjords. This forms only under very warm, damp conditions. Laterite contains aluminium oxide in addition to the iron oxide which gives it a red colour. In ordinary interbasaltic beds, however, the aluminium is bound in clay minerals.

From the *West Fjords.* Size: *10 x 15 cm.*

FOSSIL LEAVES

DESCRIPTION: Fossil leaves have long been known in Iceland. These are remnants of leaves which have been shed by trees, and then carried by the wind or water to places where they have been preserved in claystone beds or in fine-grained sandstone. Various other plant remnants are found in such layers, among them fruits, needles and even cones of evergreens, as well as microscopic grains of pollen.

OCCURRENCE: The best-known locations for leaf fossils from the Tertiary are in the West Fjords. The most famous is Surtarbrandsgil (= Lignite Gully) north of Brjánslækur. This is a protected natural site. Leaf fossils from the Ice Age are found at various places, e.g. near Svínafell in Öræfi in the southeast, and Bakkabrúnir in Víðidalur in the north.

From *Tjörnes, Northeast Iceland.* Size: *11 x 14 cm.*

FOSSIL SHELLS

DESCRIPTION: Shells have been preserved in Iceland in marine sedimentary rock that has formed near the shore, and at some depth offshore. The conditions of formation may be deduced from the grain size, layering and texture of the sedimentary rock, which is normally claystone or sandstone. The sandstone is usually rich in basaltic glass, and hence brownish in colour.

OCCURRENCE: The Pliocene Tjörnes layers in the northeast abound in shells of many different species. Layers containing shells from lower Pleistocene are found above the Tjörnes layers at Breiðavík. Layers containing shells of similar age have been found on the Snæfellsnes peninsula. Xenolithic fragments of sedimentary rock containing shells are found in hyaloclastite and tuff in Mýrdalur in the south. The oldest of these are from the end of the Tertiary. Shells are common in more-or-less consolidated sediments from the end of Pleistocene, in areas which were below sea level at that time.

Lignite. From *Skálanesbjörg, East Iceland.* Size: 7.5 x 8 cm.

LIGNITE

DESCRIPTION: Compressed plant remains are known collectively as lignite (Icelandic *surtarbrandur*). These have formed in the same way as peat in wetland regions. Lignite occurs in thin beds, rarely more than a metre in thickness. It always contains layers of ash which make it a rather poor fuel. Lignite is black or dark brown, with a dull greasy lustre. Layers of claystone that accompany lignite often share its black colour, due to oil from the lignite seeping into the clay. This is referred to as bituminous claystone. The oil may also collect as droplets in voids and fissures in the rock some distance from its place of origin.

OCCURRENCE: Lignite occurs in many places in sedimentary beds in the basalt formation. In the west it is found at the outer parts of Ísafjarðardjúp, in Arnarfjörður, Barðaströnd, Steingrímsfjörður, and above Hreðavatn. In the north and

east it is less common, but occurs e.g. at Siglufjörður, Tjörnes, Vopnafjörður and Reyðarfjörður. In the 20th century lignite was mined for fuel in wartime when there was a shortage of imported coal. Mine workings still exist at several places, best preserved at Stálfjall, Barðaströnd, Northwest Iceland.

VARIANTS: *Fossil wood* (Icel. *viðarbrandur*) consists of compressed tree trunks. It is found in among layers of lignite or related sedimentary beds. These show a clear wooden texture. *Petrified lignite* (Icel. *steinbrandur*) consists of small plant remains which have been compressed and mixed to some degree with fine-grained non-organic materials. It is greyish-black and matte, and cleaves into thin, brittle flakes. Petrified lignite of near-pure organic composition may also occur. It is black and layered with a greasy lustre.

84

Bitumen as an amygdule in basalt.
From *Hoffellslambatungur, Southeast Iceland.* Size: *9.4 x 10.5 cm.*

Flaky lignite.
From *North Iceland.* Size: *width 8.5 cm.*

Fossil wood with clear annual growth rings. From *Dýrafjörður, Northwest Iceland.* Size: *6 x 14 cm.*

From *Skagafjörður, North Iceland*. Size: *11 x 9 cm.*

SILICIFIED WOOD

DESCRIPTION: Silicified wood usually consists of jasper. Organic matter has been replaced by silica (SiO_2), and the wooden texture, and annual growth rings, are easily distinguishable.

OCCURRENCE: Silicified wood is most likely to be found among rhyolitic tephra, especially in layers of ignimbrite. It has been found, for instance, in Reykhólasveit in the west, Loðmundarfjörður in the east, Skagafjörður and Glerárdalur in the north and Drápuhlíðarfjall on the Snæfellsnes peninsula.

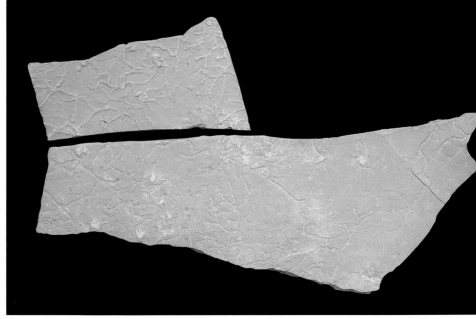

Crawl-trails (?) in claystone. From *Þórisdalur, Southeast Iceland*. Size: *maximum length 24 cm.*

LIFE MARKS

DESCRIPTION: Sometimes marks of vegetation and organisms may be seen in claystone and sandstone, and on fossils. These may be moulds of tree trunks in lava, footprints, crawl-trails, faecal pellets, imprints or casts of shells which have themselves been dissolved, leaves damaged by insects, and holes in shells caused by predator snails.

OCCURRENCE: Moulds of tree trunks in lava are common in the West Fjords, e.g. at the coast inland from Patreksfjörður (Geirseyri), at Hestur in Önundarfjörður and Hreggsstaðanúpur on Barðaströnd. A well-known site in the north is in Kotagil, Norðurárdalur in Skagafjörður. The Tjörnes layers in the northeast contain life marks of various kinds. Inland from Þórisdalur, Lón is a well-known occurrence of claystone with marks resembling crawl-trails. However, it is not known what creature left these tracks.

ZEOLITES AND RELATED MINERALS

Zeolites are a category of hydrous secondary minerals composed of sodium, potassium and/or calcium-aluminium silicates. In dry air and when warmed slightly, they lose water, and reabsorb it under wet conditions.

The word zeolite is derived from ancient Greek, from *zein* (= to boil) and *lithos* (= rock). The name refers to one of the characteristics of zeolites, that when heated they lose their water content and appear to "boil". The Icelandic name *geislasteinar* (ray stones) refers to the radiating fibres, which are typical of many common zeolites. They are soluble in hot hydrochloric acid, and when cooled gelatinous silica is deposited. Zeolites are generally white or colourless, but sometimes tinted red or yellow by minor impurities. Hardness is 2 to $5\frac{1}{2}$. Cleavage is variable, and fracture is uneven or conchoidal. Lustre is usually vitreous or pearly. The finest zeolites are those from cavities and fissures that have not been completely filled.

Fibrous/ Acicular	Tabular/ Prismatic	Granular
Scolecite	Stilbite	Chabazite
Mesolite	Heulandite	Analcime
(Natrolite)	Epistilbite	(Wairakite)
Mordenite	Levyn	Phillipsite
Thomsonite	Yugawaralite	
Laumontite	Erionite	
Phillipsite	Cowlesite	
Garronite	Thomsonite	
Gismondine	Laumontite	
Erionite		

There are 48 recognized species of zeolite. Iceland is famous for its zeolites, and about 20 of them are found there. The largest specimens of zeolites are from India and Brazil.

Zeolites are often classified by shape in three main categories, fibrous/acicular, tabular/prismatic and granular. Some are variable in shape, such as phillipsite, erionite and thomsonite. Fibrous zeolites often form sheaf-like aggregates.

Zeolites are formed during alteration and precipitation in vugs and vesicles. As water percolates down into a rock pile, it heats up, depending on the depth of penetration. The highest temperatures are reached in the vicinity of intrusive bodies. The hot water dissolves various substances from the rock, and new so-called secondary or alteration minerals are formed. On cooling, the minerals precipitate to form amygdules and fissure fillings. It depends on the temperature and chemical composition of the solution which minerals are formed. Zeolites form at relatively low temperatures. They are not found where alteration of rocks is most intense, with temperatures exceeding 230°C, such as in high-temperature geothermal areas.

Various minerals besides quartz and calcite commonly occur with zeolites. Most are hydrated silicates, while others are sulphates. These minerals, which include apophyllite, gyrolite and okenite, are often classified with zeolites although they do not strictly belong there, as they do not contain aluminium.

From *Teigarhorn, East Iceland*. Size: *24 x 15 cm.*

SCOLECITE

Crystal system: monoclinic
Hardness: 5
Specific gravity: 2.25–2.31

Cleavage: perfect in one direction
parallel to longitudinal axis
$CaAl_2Si_3O_{10} \cdot 3H_2O$

DESCRIPTION: Scolecite is a typical fibrous zeolite. Four-sided, often rather densely packed, flattened, fibrous crystals form groups or aggregates. The crystal fibres radiate from a single point, growing thicker, and may separate, towards an obtuse point at the end. Thicker scolecite fibres may have faint longitudinal marks. Scolecite is colourless or white, with vitreous or slightly silky lustre. Fracture is uneven. Longitudinal cleavage shows slight cleavage cracks at right angles to direction of crystal fibres. Crystal fibres are commonly 1–3 cm, but may be longer, up to 10 cm. When scolecite fills small amygdules the crystals are correspondingly small and the radiating pattern is scarcely visible.

OCCURRENCE: Scolecite is common in olivine basalt from the Tertiary. It is found with various other zeolites that are relatively low in silica, such as mesolite, chabazite, thomsonite and analcime, but occurs at greater depths in the rock strata than the last three of these. Scolecite and mesolite are typical for the lower part of the exposed basalt formation, and one of the zeolite zones is named after it, the mesolite-scolecite zone.

NAME: The word *scolecite* was coined in 1813. It derives from the Ancient Greek *skolex* (= worm), a reference to the way that scolecite "curls" like a worm when heated in a flame.

End faces of the crystal needles are clearly discernible. From *Reyðarfjörður, East Iceland*. Size: *5.5 x 11 cm*.

Common form of scolecite. Fibrous crystal aggregates radiating from several points. From *Vestrahorn, Southeast Iceland*. Size: *9 x 6 cm*.

Crystal aggregate showing faint cracks at right angles to the fibres. From the *East Fjords*. Size: *11 x 7 cm*.

Delicate fibres of uniform size radiate from a single point.
From *Southeast Iceland.* Size of crystal spray: *2.3 x 1.5 cm.*

MESOLITE

Crystal type: monoclinic
Hardness: 5
Specific gravity: 2.25–2.26

Cleavage: Perfect in two planes at an angle to the longitudinal axis.

$Na_2Ca_2Al_6Si_9O_{30} \cdot 8H_2O$

DESCRIPTION: Mesolite forms small-scale fibrous radiating clusters or tufts. The crystals are slender fibres, whose longitudinal faces can hardly be identified using a magnifying glass. The crystal fibres, rarely more than 3 cm long, are of uniform width. They separate at the ends, forming a spiky surface. The fibres are brittle, and break easily if not handled with care. Mesolite is usually white, translucent or greyish, but may be tinted by impurities, usually pink or red. It has a vitreous or silky lustre. Fracture is uneven, if it can be identified.

OCCURRENCE: Mesolite commonly appears with scolecite in olivine basalt from the Tertiary. It occurs with other zeolites relatively low in silica, and is the second most common mineral in the mesolite-scolecite zone of the basalt formation.

NAME: The word *mesolite*, coined in 1816, is drawn from the ancient Greek *mesos* (= in the middle), a reference to the fact that when first chemically analysed its composition was found to be between scolecite and natrolite.

Natrolite resembles mesolite and scolecite. It has not been found independently in Iceland, but it was identified (in Skriðdalur, East Iceland) as a "growth" on mesolite, where it grew on the end of mesolite's fibrous crystals. It is white or colourless, sometimes grey or yellowish. The name *natrolite*, coined in 1803, derives from *natrium* (= sodium).

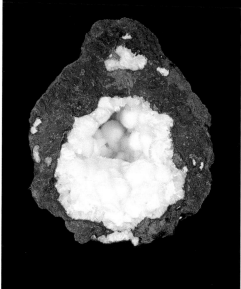

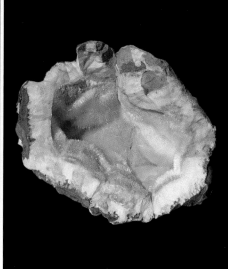

Crystal clusters in hemispherical forms with a spiky surface due to projecting crystal ends. From the *East Fjords*. Size: *5.5 x 6.5 cm*.

Pink variant. From *Skagafjörður, North Iceland*. Size: *10 x 7.5 cm*.

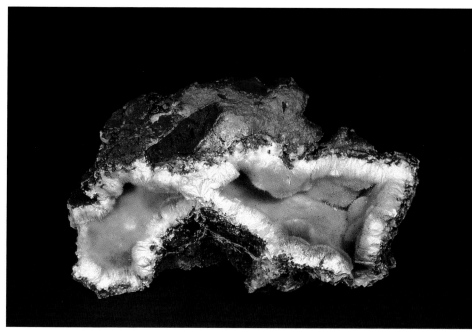

The reddish colour is due to iron oxides. From *Skagafjörður, North Iceland*. Size: *length 24 cm*.

Crystal tuft resembles a lock of hair. From *Hvalfjörður, West Iceland*. Size: *length of tuft 2.5 cm.*

MORDENITE

Crystal type: Orthorhombic
Hardness: 4–5
Specific gravity: 2.12–2.15

Cleavage: Perfect, but indistinct due to small size
$(Na_2K_2Ca)Al_2Si_{10}O_{24} \cdot 7H_2O$

DESCRIPTION: Mordenite forms delicate flexible fibres and tufts which are yielding to the touch. It often forms a dense mass beneath the tufts and in filled cavities. It rarely exceeds 0.5 cm in length, although fibres up to 2 cm long have been found. Mordenite resembles a tuft of cotton wool or a film of mould, often with silky lustre. Where wetted the mordenite tufts become compressed and tattered. Mordenite is generally white, but sometimes reddish-brown. The crystals of mordenite being minute, it must be magnified more than 100 times in order to identify the crystal form.

OCCURRENCE: Mordenite is one of the more common zeolites. It is generally found in tholeiite and silica-rich rock but not normally in olivine basalt. It occurs in the lower part of the exposed lava pile. It often occurs with chalcedony, quartz and epistilbite, but generally occurs as a sole mineral in the vesicles.

NAME: The name *mordenite*, coined in 1864, is drawn from the locality *Morden* in Nova Scotia, Canada.

Flokite is an obsolete name for mordenite drawn from Hrafna-*Flóki*, who sailed to Iceland before its settlement.

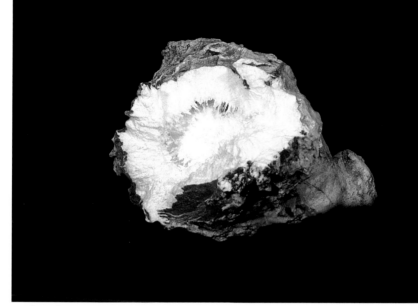

Mordenite forming a dense mass on the wall of the cavity, but leaving a space in the centre.
From *Þorskafjarðarheiði, West Fjords.* Size: *10 x 7.5 cm.*

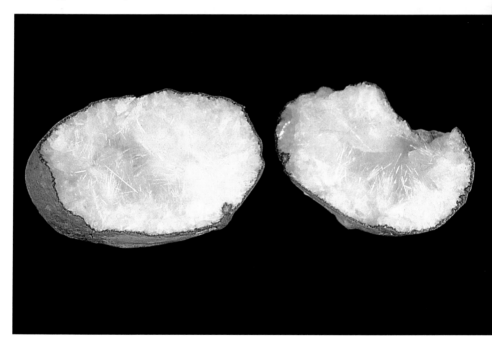

Delicate crystal fibres forming irregular sprays.
From *Skarðsheiði, West Iceland.* Size: *diameter of amygdule 4 cm.*

From *West Iceland*. Size: *amygdule 5 x 12 cm.*

THOMSONITE

Crystal type: orthorhombic
Hardness: 5
Specific gravity: 2.25–2.44

Cleavage: Perfect, in one direction
$NaCa_2Al_5Si_5O_{10} \cdot 6H_2O$

DESCRIPTION: Thomsonite is a radiating zeolite. Crystals are elongated, slightly flattened, rather thin fibres, with a slanted end. They form dense masses of radiating clusters, with mamillary structures whose surface is finely bristled. The cleavage surface often reveals concentric stripes of differing tone: white, translucent, pale blue. Otherwise its colour is mainly milky white, while reddish and brown thomsonite has also been found. Fracture is uneven or indistinctly conchoidal. Thomsonite is translucent with vitreous or pearly lustre. Crystals are most commonly less than 1 cm in length, but greater lengths also occur.

OCCURRENCE: Thomsonite is very common in olivine basalt, often occurring with chabazite, levyne, phillipsite and calcite. The uppermost zeolite zone of the Icelandic basalt formation is called chabazite-thomsonite zone from its most common minerals.

NAME: *Thomsonite*, coined in 1820, is drawn from the name of chemist Th. *Thomson*. An obsolete name for thomsonite is *farolite*, named after the *Faroe* Islands.

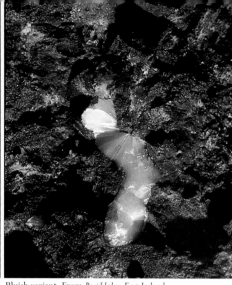

The large hemispherical aggregates are thomsonite, while the small spiky spherules are okenite. From *Hvalfjörður, West Iceland.* Size: *aggregates 8 mm.*

Bluish variant. From *Breiðdalur, East Iceland.* Size: *1.4 x 0.4 cm.*

Reddish-brown variant. From *Skarðsheiði, West Iceland.* Size: *3.7 x 4 cm.*

Small prisms project from the surface of the aggregate. From *Esja, Southwest Iceland.* Size: *aggregates 5 mm.*

Crystal fibres and tabular crystals with oblique ends. From *West Iceland*. Size: *crystal fibre 1.5 cm.*

LAUMONTITE

Crystal type: monoclinic
Hardness: 3–3½
Specific gravity: 2.20–2.41

Cleavage: two perfect longitudinal
cleavages
$CaAl_2Si_4O_{12} \cdot 4H_2O$

DESCRIPTION: Laumontite is classified either as a radiating or tabular zeolite. It forms thin, elongated fibres or prisms, with longitudinal marks and a square end. A cross-section of the crystals is square. It is normally white, but sometimes pale pink or reddish-brown. Common length is 0.5 cm, but longer fibres are also widely found. It may also occur as radiating crystals. Cleavage is perfect, with a slight pearly lustre on the broad cleavage surface. On giving off its water upon drying laumontite becomes very brittle. The finest specimens are therefore found in newly opened amygdules in nature. The fragile crystals are rather loosely distributed in the cavities, but they often occur in large quantity, and tend to crumble when the rock is broken. Laumontite deteriorates when stored in dry conditions, becoming dull and chalky. Hence it is difficult to preserve good samples.

OCCURRENCE: Laumontite is the zeolite that forms at the highest temperature, up to 230°C. It is common in deeply eroded basalt formations but occurs also at higher levels in surroundings of high-temperature geothermal alteration. It is normally found as a sole mineral in cavities, or with calcite. The lowest zeolite zone is named the laumontite zone after this mineral.

NAME: *Laumontite* was named in 1808 after Gillet de *Laumont*, a Frenchman who discovered the first samples that were identified.

Tabular crystals in a cluster. From the *East Fjords.* Size: *10 x 7 cm.*

Laumontite sometimes forms radiating sheaves. From *Hjallaháls, West Fjords.* Size: *9 x 6 cm.*

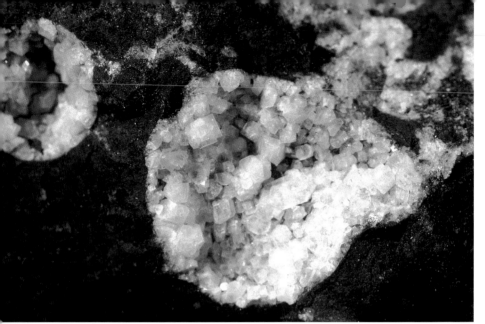

Translucent, blocky crystals with typical crystal faces. From *Svarfaðardalur, North Iceland.* Size: *crystals 2–5 mm.*

PHILLIPSITE

Crystal type: monoclinic
Hardness: 4½
Specific gravity: 2.20

Cleavage: distinct in one direction,
otherwise indistinct
$(K,Na,Ca)_{1-2}(Si,Al)_8O_{16}·6H_2O$

DESCRIPTION: Phillipsite is colourless or white, but may turn pale brown upon weathering. It has vitreous lustre, is translucent and generally forms very small crystals, although examples 0.5 cm long have been found. Crystals are pseudotetragonal, rather thick, with oblique end surfaces forming an obtuse point. Phillipsite occurs both as twinned individual crystals, often cruciform due to penetration twinning, and in radiating clusters of fibrous aggregates. The fibres are fairly broad and laminar with a grooved pattern, and oblique end faces. Phillipsite is counted as a common mineral in Iceland, but it is not easily identifiable,

due to its small size. It resembles thomsonite, scolecite and stilbite when radiating, but is similar to apophyllite when columnar, in which case it lacks perfect cleavage.

OCCURRENCE: Phillipsite is found mainly in olivine basalt in the upper parts of the lava pile, where it occurs with chabazite, levyne and thomsonite, though not necessarily in the same cavities.

NAME: *Phillipsite* was named in 1825, after English mineralogist W. *Phillips* (1775–1829). Phillipsite found in the West Fjords in the mid-19th century was named *christianite* after King *Christian* VII of Denmark and Iceland.

Cruciform twinning. From *Esja, Southwest Iceland.*
Size: *cross 3–4 mm. (Photo SSJo).*

Blocky phillipsite on thomsonite.
From *Eilifsdalur, Southwest Iceland.* Size: *crystals c. 1 mm.*

Mamillary clusters of crystal sheaves.
From *West Iceland.* Size: *4 x 5 cm.*

Whitish equant crystals on chabazite.
From *Esja, Southwest Iceland.* Size: *crystals 1–2 mm.*

101

White sheaves of tabular crystal clusters on chabazite. From a road tunnel in the *West Fjords*. Size: *crystals 5—7 mm.*

GISMONDINE

Crystal type: monoclinic
Hardness: 4½
Specific gravity: 2.12–2.28

Cleavage: indistinct
$Ca_2Al_4Si_4O_{16}\cdot 9H_2O$

DESCRIPTION: Gismondine is colourless or white, generally radiating and of small size. It usually forms matte half-spheres with a grooved surface. Gismondine also forms individual crystals in a pseudotetragonal (dipyramidal) form, rarely more than 0.5 cm in length. Fracture is conchoidal. Gismondine is related to phillipsite and garronite, and is found in the same environment.

OCCURRENCE: Gismondine is rare, found principally in cumulate type porphyritic basalt and olivine basalt in the lower part of the chabazite-thomsonite zone, together with other zeolites relatively low in silica, such as chabazite, thomsonite, phillipsite and levyne, and generally in the same cavities.

NAME: *Gismondine* was named in 1817 after Italian mineralogist Carlo G. *Gismondi*, who first described it.

Spheres with radiating interior. Crystal faces project from the surface of the spheres.
From a road tunnel in the *West Fjords*. Size: *crystals 5—7 mm.*

Spheres with typical twinned tabular crystals.
Sometimes likened to the form of the Sydney Opera House.
From *Barðaströnd, West Iceland*. Size: *crystals 4—5 mm (Photo SSJo).*

Gismondine spheres comprising tabular
crystals, growing on chabazite.
From *Dalir, West Iceland*. Size: *crystals 8 mm.*

Rounded clusters and typical fracture. From *Esja, Southwest Iceland.* Size: *0.9 x 1.9 cm.*

GARRONITE

Crystal type: orthorhombic
Hardness: 4½
Specific gravity: 2.13–2.18

Cleavage: perfect
$Na_2Ca_5Al_{12}Si_{20}O_{64}·27H_2O$

DESCRIPTION: Garronite forms radiating, densely packed crystal clusters, in which the radiating form is rather indistinct. The crystal clusters are semi-spherical, typified by concentric fracture surfaces at right angles to the fibres. Garronite is milky-white or colourless with vitreous lustre. It often forms amygdules about 1 cm in diameter, which frequently fill the entire cavity. Garronite is fairly rare. It is closely related to phillipsite. In half-filled cavities, the garronite amygdules are often overgrown with phillipsite.

OCCURRENCE: Garronite occurs in olivine basalt in the chabazite-thomsonite zone and in the analcime zone. It is found together with other zeolites relatively low in silica, such as chabazite, thomsonite (which it may resemble) and levyne. In east Iceland garronite is well known in the extensive Grænavatn porphyritic basalt group at 700 to 900 metres above sea level. Garronite is fairly rare, but has been found in various other places in recent years, e.g. in the West Fjords.

NAME: *Garronite* was named in 1962 by the geologist G.P.L. Walker, after the place where it was first found, *Garron* in Northern Ireland.

Fracture surfaces showing concentric curves and radial cracks. From *Norðurárdalur in Borgarfjörður, West Iceland.* Size: *radius of infilling 1 cm.*

Vesicular basalt with garronite amygdules. From *Hvalfjörður, West Iceland.* Size: *largest cavity 7 mm.*

Minute greyish spheres growing on analcime. From *Hvalfjörður, West Iceland*. Size: *1.7 x 2.3 cm*.

COWLESITE

Crystal type: monoclinic
Hardness: 2
Specific gravity: 2.12–2.28

Cleavage: perfect
$CaAl_2Si_2O_8 \cdot 4H_2O$

DESCRIPTION: Cowlesite forms grey or white, very small, thin laminar crystals with a pointed end, commonly one mm or less in length. Lustre is vitreous or pearly. Cowlesite generally forms small spherical clusters. It may resemble thomsonite, but the two may be distinguished by external appearance and hardness. Cowlesite clusters have a rougher surface than thomsonite due to the pointed fibres, and cowlesite is also much softer.

OCCURRENCE: Cowlesite occurs in olivine basalt, generally with levyne, to which it is related, and also with analcime, thomsonite and heulandite. It is not found in tholeiitic basalt or acid rock. It is rare in Iceland, and only a few locations have been discovered, the best-known of which are Mjóadalsá in Borgarfjörður (where it was first found in Iceland) and Álftafjörður, both in the west. Cowlesite is found in the chabazite-thomsonite zone, and down to the mesolite-scolecite zone.

NAME: *Cowlesite* was named in 1975 in honour of John *Cowles*, an American mineral collector.

The radiating form of spheres, and the rough surface, are visible under great magnification. From *Álftafjörður, Snæfellsnes Peninsula, West Iceland*. Size: *spheres less than 1 mm (Photo SSJo)*.

Small greyish spheres of cowlesite. From *Eilifsdalur near Hvalfjörður, West Iceland*. Size of cavity: *8 x 12 mm*.

Loosely intergrown tabular crystal bundles. From *Borgarfjörður, West Iceland*. Size: *14.5 x 7.5 cm.*

STILBITE

Crystal type: monoclinic, orthorhombic
Hardness: 3½–4
Specific gravity: 2.12–2.21

Cleavage: perfect
$NaCa_2Al_5Si_{13}O_{36} \cdot 14H_2O$

DESCRIPTION: Stilbite appears in many forms, but most commonly as thick, tabular crystals with pointed terminations. Crystals often grow wider towards the end, forming sheaf-like aggregates. Sheaves may form at both ends in a "bow-tie". Stilbite also occurs as clusters of elongated six-faced crystals with edge-like termination. A radiating variant is also common. This has laminar crystals, ending in slightly offset smooth, oval surfaces. Stilbite is generally milky-white, but clear and translucent variants exist, and also coloured types, usually reddish-brown but sometimes greenish. Stilbite has vitreous lustre, especially on the domed or oval ends of crystal bundles, while the broad face of the crystal has pearly lustre. Fracture is uneven. Crystals are commonly 1–2 cm, although 5–10 cm crystals have been found.

OCCURRENCE: Stilbite is one of the commonest zeolites. It is found in amygdaloidal basalt in the analcime zone and below. It is most abundant in tholeiite low down in the lava pile, especially in the East Fjords (Berufjörður). It is a common amygdule in geothermal systems.

NAME: *Stilbite* was named in 1801. The name derives from the Ancient Greek *stilbein* (= glow), in reference to the pearly lustre on the broad face of the crystal. *Desmine* (= sheaf) is a synonym for stilbite sometimes used in German and Danish.

VARIANTS: *Stellerite* is an orthorhombic calcium-rich variant of stilbite, generally with smooth, oval, lustrous end faces. It has been found in Iceland, but is rare. It was named in 1909 in honour of G.W. *Steller*.

Common form. The crystal aggregates resemble sheaves of wheat. From *Southeast Iceland*. Size: *23 x 35 cm*.

Irregularly distributed tabular crystals with clear faces. From *Borgarfjörður, West Iceland*. Size: *crystals 7.5 x 3 mm*.

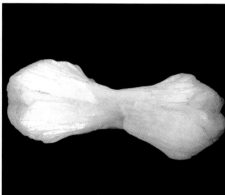

Dense crystal aggregates. The broad side of the crystals showing pearly lustre. From the *East Fjords*. Size: *8 x 6 cm*.

Bow-tie shaped variant is found mainly in the East. From the *East Fjords*. Size: *4.3 x 0.7 cm*.

Crystal clusters with rounded ends and vitreous lustre on crystal faces. From *Hvalfjörður, West Iceland*. Size: *8 x 10 cm*.

Bow-tie shaped variant. Individual bundles are clearly distinguishable. From the *East Fjords*. Size: *6 x 7 cm*.

From *Teigarhorn, East Iceland.*

HEULANDITE

Crystal type: monoclinic
Hardness: 3½–4
Specific gravity: 2.1–2.29

Cleavage: perfect
$(Na,Ca)_{2-3}Al_3(AlSi)_2Si_{13}O_{36}\cdot12H_2O$

DESCRIPTION: Heulandite forms tabular crystals. It is translucent, or even transparent in newly opened cavities, rarely reddish. It generally forms thickish trapezoid aggregate or clusters with truncated points. The clusters are sometimes slightly curved, or even almost semi-circular in form. The crystal form is sometimes referred to as coffin-shaped. Heulandite cleaves easily, with strong pearly lustre on cleavage surfaces. Fracture is uneven. Common size is 0.5 to 2 cm, but examples up to 10 cm have been found in Iceland. Heulandite is easily identified by perfect cleavage and pearly lustre. It resembles stilbite.

OCCURRENCE: Heulandite is one of the commonest zeolites, often found with stilbite, especially in tholeiite, downwards from the analcime zone. It is most common in lower parts of the lava pile. The largest and finest examples are found in the East Fjords.

NAME: *Heulandite* was named in 1822 after British mineral collector H. *Heuland.* Heulandite was first identified as a separate mineral in a sample from Teigarhorn, East Iceland, collected by Heuland.

VARIANTS: *Clinoptilolite* is a silica-rich variant of heulandite, whose crystals are strongly curved.

Red variant with curved crystal faces.
From the *East Fjords.* Size: *heulandite 4 x 2.3 cm.*

Tabular crystals with characteristic crystal faces and angles. From *Teigarhorn, East Iceland.*
Size: *crystal 1.2 x 1.6 cm.*

Stacks of blocky tabular heulandite crystals. From *Teigarhorn, East Iceland.* Size: *9 x 5 cm.*

111

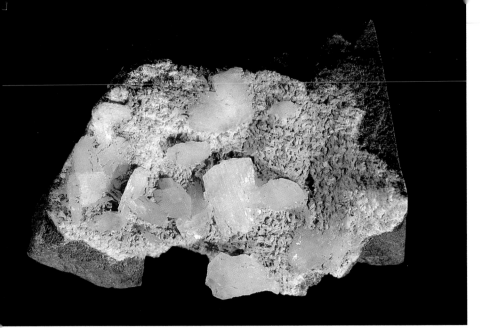

Pink variant. From *Skarðsheiði, West Iceland*. Size: *3.1 x 4.4 cm.*

EPISTILBITE

Crystal type: monoclinic
Hardness 4–4½
Specific gravity: 2.22–2.68

Cleavage: perfect
$CaAl_2Si_6O_{16} \cdot 5H_2O$

DESCRIPTION: Epistilbite forms crystals which are almost triangular in cross-section, elongated, rather thick, white or colourless, sometimes slightly reddish or greenish. They taper towards the edges, which are truncated at the very margin. Twinning is common, forming a rhomboid cross-section. Crystal size is commonly 0.5–1 cm, but crystals of up to 3 cm have been found in Iceland. Cleavage is perfect, both longitudinally and at an angle to the longitudinal axis, producing a dome with non-parallel crystal faces on the ends. Lustre is vitreous or slightly pearly, in which case it resembles stilbite. Crystals often occur as dense aggregates or clusters. It is distinguishable from stilbite by triangular slightly oblique crystal ends.

OCCURRENCE: Epistilbite is fairly rare. It is found mainly in tholeiite in the lower parts of the lava pile. It is mainly found in the East Fjords, but also in Hvalfjörður and Borgarfjörður in the west.

NAME: *Epistilbite* was named in 1826, in reference to its likeness to stilbite (Greek *epi* = after), although the two are unrelated.

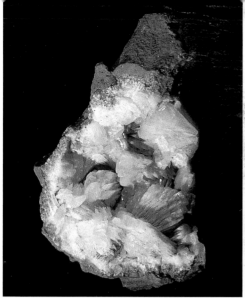

Cluster of crystals with rhomboid end faces. From *Teigarhorn, East Iceland*. Size: *4.5 x 4 cm.*

Twinned crystals, with rhomboid cross-section. Blocky aggregates of tabular crystals. From *Skarðsheiði, West Iceland*. Size: *3 x 2 cm (Photo SSJo).*

Sheaf with slightly tabular crystals. From *Húsafell, West Iceland*. Size: *crystal sheaf 1.3 x 1 cm.*

113

Hexagonal plates typically stand on edge. From *Gjörvidalur, West Fjords*. Size: *diameter of cavity 2 cm.*

LEVYNE

Crystal type: hexagonal
Hardness 4–4½
Specific gravity: 2.09–2.16

Cleavage: indistinct
$(Ca,Na_2,K_2)Al_2Si_4O_{12}\cdot 6H_2O$

DESCRIPTION: Levyne forms colourless or white, thin platy crystals with hexagonal outline and grooves at each corner. Lustre is vitreous. The size of crystals is usually several millimetres. Levyne is best identified where crystals are irregularly spaced standing on edge in a cavity. Twinning is common along the broad faces. The grooves which are characteristic on the edges result from the individual crystal plates being slightly rotated relating to each other.

OCCURRENCE: Levyne is fairly common in olivine basalt, especially in the chabazite-thomsonite zone of the basalt formation, extending down to the mesolite-scolecite zone. It is generally found as the sole mineral in cavities, but other types may be found nearby, e.g. gismondine, garronite and phillipsite.

NAME: *Levyne* was named in 1825 after French mineralogist A. *Lévy*.

114

Transparent hexagonal plates. From *Hvalfjörður, West Iceland.*
Size: *crystals about 3 mm.*

Thin platy crystals with twinning. Grooves are
clearly visible at corners. The grey external coating
is intergrown erionite and offretite. From *Barðaströnd,
West Iceland.* Size: *diameter of cavity 1 cm (Photo SSJo).*

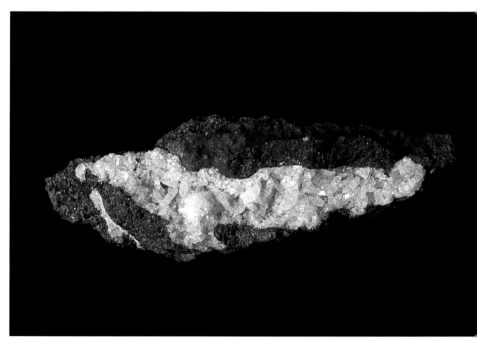

Levyne crystals with hexagonal outline on edge in a cavity. Grooves are visible at corners.
From *Eilífsdalur, near Hvalfjörður, West Iceland.* Size: *width of cavity about 1 cm.*

115

From *Hvalfjörður, West Iceland.* Size: *5.5 x 11 cm.*

YUGAWARALITE

Crystal type: monoclinic
Hardness: 4½
Specific gravity: 2.19–2.25

Cleavage: distinct in two directions
$CaAl_2Si_6O_{16} \cdot 4H_2O$

DESCRIPTION: Yugawaralite forms colourless or sometimes white platy crystals with oblique or truncated ends. Crystals are normally small, but may exceed one cm in size, in which case they are elongated tabular crystals in irregular brittle clusters. It has vitreous lustre.

OCCURRENCE: Yugawaralite is rare, and has only been found in a few places in Iceland. It occurs in highly altered andesite and tholeiite at the margins of extinct geothermal systems. The finest Icelandic examples of yugawaralite have been found in Hvalfjörður in the west. It occurs with calcite, quartz, and zeolites such as heulandite and stilbite.

NAME: *Yugawaralite* was named in 1952 after the *Yugawara* geothermal area in Japan, where this mineral was first discovered.

Tabular crystals with characteristic crystal faces and corners. From *Hvalfjörður, West Iceland*. Size: *crystals 3–5 mm (Photo SSJo)*.

Yugawaralite as fissure lining. From *Hvalfjörður, West Iceland*. Size: *17.5 x 6.5 cm.*

Curly fibres resembling wool. From *Skagafjörður, North Iceland*. Size: *c. 4 cm.*

ERIONITE

Crystal type: hexagonal
Hardness: 4
Specific gravity: 2.02–2.13

Cleavage: indistinct
$(K_2,Ca,Na_2)_2Al_4Si_{14}O_{36}\cdot 15H_2O$

DESCRIPTION: Erionite forms filaments, generally in dense aggregates. The crystals form undulant flat fibres, which are flexible and yielding. Erionite is white with pearly lustre. Crystal aggregates are often 2–3 cm long, resembling curly hair. Tufts of erionite are soft and flexible, reminiscent of frayed rope, when plucked from cavities in the rock. Erionite also occurs in Iceland as small six-sided needle-shaped crystals.

OCCURRENCE: Erionite is rare. It occurs in tholeiite and olivine basalt low down in the lava pile. It occurs on the Tröllaskagi peninsula in the north, in the East Fjords and in Hvalfjörður in the west.

NAME: *Erionite* was named in 1898, with reference to its wool-like appearance. The Ancient Greek *erion* means wool.

VARIANTS: *Offretite* is filamentous like erionite. It occurs with erionite, growing on levyne, from its laminar crystal surfaces. It has recently been found in Iceland. The name, coined in 1890, is in honour of a French professor, Albert J. *Offret*.

118

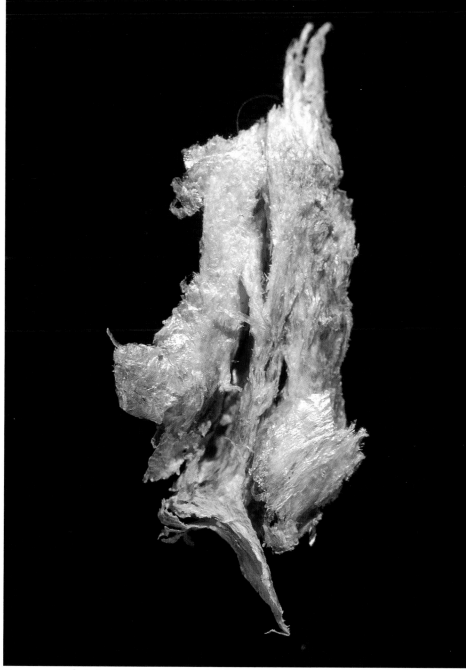

Dense aggregate of erionite, frayed at sides and ends where filament form can be seen.
From *Svarfaðardalur, North Iceland*. Size: *length 4.5 cm.*

From *Súgandafjörður, West Fjords*. Size: *4 x 5.5 cm.*

CHABAZITE

Crystal form: triclinic
Hardness: 4½
Specific gravity: 1.97–2.20

Cleavage: distinct in three directions
$CaAl_2Si_4O_{12} \cdot 6H_2O$

DESCRIPTION: Chabazite forms white or clear, rarely yellowish brown or slightly reddish, near-cubic crystals, often with rhomboid faces. Crystals are often twinned, penetrating each other with the corners projecting from the faces. Lustre is vitreous and fracture uneven. Common size is several millimetres, but crystals of up to 1.5 cm may be found. Chabazite occurs in a number of different forms, depending on which faces of twinned crystals are developed.

OCCURRENCE: Chabazite is the commonest zeolite in Iceland. It is typical of the uppermost zeolite zone, together with thomsonite, and is mainly found in

olivine basalt. Chabazite is not found exclusively in the zeolite zone of the same name, but also lower down in the lava pile, where the finest examples occur.

NAME: *Chabazite*, coined in 1792, is from the Ancient Greek *chabazios* (= hail). It is mentioned in the Ancient Greek poem *Peri lithon* (On Stones) by Orpheus.

VARIANTS: *Phacolite* is a twinned form of chabazite which grows from a hexagonal base in low pyramidal form with wedge-shaped grooves and stepped sides. *Phacolite* means "lenticular stone".

120

Common twinned form.
From *West Iceland*. Size: *8.3 x 8.3 cm.*

Twinned pyramidal form.
From *Esja, Southwest Iceland*. Size: *largest crystals 2 mm.*

Yellow variant. From *Skarðsheiði, Borgarfjörður, West Iceland*.
Size: *largest crystals 2 mm.*

Phacolite variant.
From *Esja, Southwest Iceland*. Size: *largest crystal 1.3 cm.*

From *Oddsskarð, East Fjords.* Size: 3 x 1.7 cm.

ANALCIME

Crystal type: cubic
Hardness 5–5½
Specific gravity: 2.22–2.63

Cleavage: indistinct
$NaAlSi_2O_6 \cdot H_2O$

DESCRIPTION: Analcime (or analcite) forms colourless or white many-sided (trapezohedral) crystals with vitreous lustre. It is almost impossible to cleave, and fracture is conchoidal. It occurs either as individual crystals or as clusters which glitter, in the case of the colourless crystals. Transparent crystals may appear black due to the colour of the underlying rock. Common size is 0.2–0.5 cm, rarely more than 1 cm, although crystals of up to 4 cm occur in Iceland. Analcime is easily identified by its crystal form and strong vitreous lustre.

OCCURRENCE: Analcime is common, occurring in olivine basalt. In the analcime zone of basalt formation it is found together with chabazite and thomsonite.

NAME: *Analcime*, coined in 1801, is from the Ancient Greek *analkis* (= powerless), in reference to the fact that the crystals do not produce a static charge when rubbed.

VARIANTS: *Wairakite* is a calcium-rich variant (or closely related mineral), which forms at 200–300°C, i.e. at higher temperatures than other zeolites. It is well known in Iceland, but small in scale, and has mainly been found in boreholes in high-temperature geothermal areas. The name, coined in 1955, is from the place-name *Wairakei* in New Zealand.

Crystal form is emphasised by the coating on the crystal faces. From *Hvalfjörður, West Iceland.* Size: *crystals up to 5 mm.*

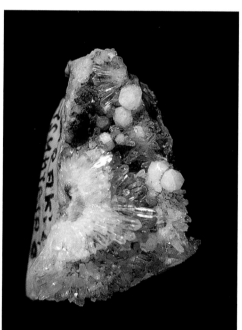

Wairakite. From *Snæfellsnes, West Iceland.*
Size: *spherules 3–4 mm.*

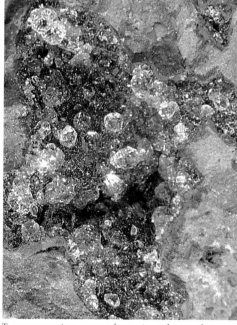

Transparent variant, commonly seen in newly opened
cavities. From the *East Fjords.* Size: *crystals up to 5 mm.*

Green variant. Four-sided pyramids. From *Hvalfjörður, West Iceland.* Size: *2.8 x 3.8 cm.*

APOPHYLLITE

Crystal type: tetragonal
Hardness: 4½–5
Specific gravity: 2.33–2.37

Cleavage: perfect in one direction
$KFCa_4Si_8O_{20} \cdot 8H_2O$

DESCRIPTION: Apophyllite forms four-sided crystals with truncated edges forming rhomboid faces that end in a pyramid. The apex is sometimes truncated, in which case the appearance is cubic. Common size is 1–2 cm, but occurs in Iceland up to twice this size. Apophyllite also occurs as tabular crystals. It cleaves easily at right angles to the columns. It has a vitreous lustre, but pearly lustre on cleavage surfaces. Fracture is uneven and conchoidal. Apophyllite is colourless or white, often greenish, sometimes yellowish or reddish. It forms either separated crystals or clusters in which the crystals are irregularly distributed. At first sight apophyllite may resemble quartz, but quartz is six-sided and much harder.

OCCURRENCE: Apophyllite is common among zeolites in olivine basalt. It occurs in all the zeolite zones except the upper parts of the chabazite-thomsonite zone. The finest examples are from the mesolite-scolecite zone.

NAME: The name of *apophyllite* is ancient, from the Greek *apo-* (= away) and *phyllon* (= leaf), in reference to the fact that it cleaves into flakes when heated in a flame. At the Uxahver hot spring in north Iceland, apophyllite grows on petrified wood. The name *Oxahverite,* long obsolete, was coined for this variety.

124

Blocky and pyramidal crystal forms. From *West Iceland*. Size: 7 x 5.5 cm.

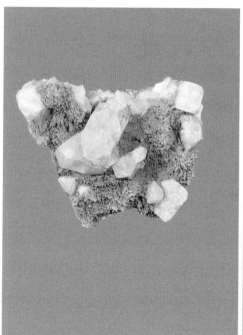

Four-sided prisms with pyramidal ends.
From *Breiðdalur, East Iceland*. Size: *crystals 1–2 cm.*

Apophyllite with truncated end surfaces, accompanied by stilbite. From *Hvalfjörður, West Iceland*. Size: *3.3 x 5.2 cm.*

125

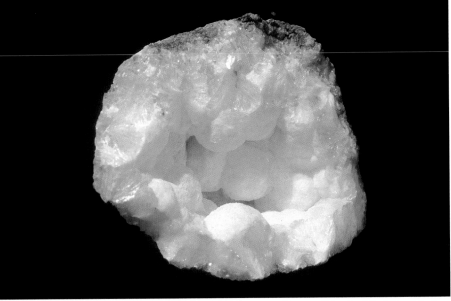

From *Breiðdalur, East Iceland*. Size: *3.5 x 3 cm.*

GYROLITE

Crystal type: trigonal
Hardness: 3–4
Specific gravity: 2.3

Cleavage: perfect in one direction
$NaCa_{16}(Si_{23}Al)O_{60}(OH)_5 \cdot 15H_2O$

DESCRIPTION: Gyrolite forms platy hexagonal crystals with irregular outlines. It is white with silky or vitreous lustre, and cleaves easily into thin flakes. Gyrolite normally grows in clusters or knobs showing a fan-shape on fracture surfaces. The appearance is somewhat similar to stilbite. The surface of the knobs is uneven and rough.

Common size is 0.5–1 cm, but knobs have been found in Iceland up to 2 cm in diameter.

OCCURRENCE: Gyrolite is fairly rare. It occurs in olivine basalt down in the mesolite-scolecite zone.

NAME: *Gyrolite* was named in 1851 with reference to the crystal form, from the Latin *gyrus* (= ring, spinning wheel).

126

Radiating aggregates show up in fracture surfaces. From *Kistufell, Esja, Southwest Iceland*. Size: *4.5 x 3.5 cm.*

Clusters and knobs of platy crystals, growing over calcite. Yellowish hue due to impurities.
From *Hvalfjörður, West Iceland*. Size: *knobs 5–8 mm.*

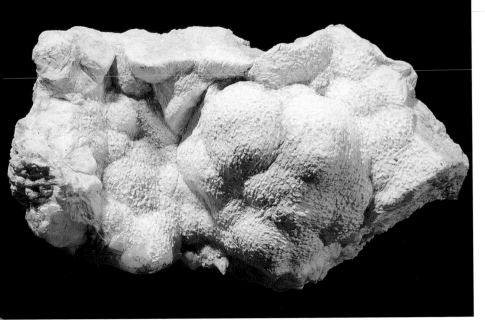

From *Svarfaðardalur, North Iceland.* Size: *12 x 7 cm.*

OKENITE

Crystal type: triclinic
Hardness: 4½–5
Specific gravity: ≈2.3

Cleavage: perfect in one direction
$Ca_{10}Si_{18}O_{46} \cdot 18H_2O$

DESCRIPTION: Okenite is white, more rarely almost colourless, with slight vitreous or pearly lustre. Crystals are filamentous, often forming a dense amygdule with indistinct radiate pattern. It also occurs as small spherules with small projecting needles, resembling a tuft of cotton wool. Common size of okenite in Iceland is several centimetres, but spherules have not been found more than a few millimetres across.

OCCURRENCE: Okenite is rare. It occurs in olivine basalt in the lower parts of the mesolite-scolecite zone and down into the laumontite zone. Okenite is generally found associated with apophyllite.

NAME: *Okenite* was named in 1828 after German naturalist L. *Oken* (d. 1851).

128

Radiating knobs with rough surface. Normally found, as here, with apophyllite.
From *Hvalfjörður, West Iceland.* Size: *width of amygdule 3 cm.*

Small bulbous aggregates of radiating fibres.
From *Borgarfjörður, West Iceland.* Size: *diameter of aggregates 2–3 mm.*

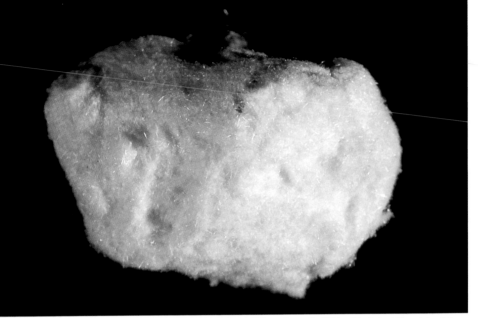

Mass of short, slender crystal fibres. From *Hvalfjörður, West Iceland*. Size: *1.1. x 1.6 cm.*

THAUMASITE

Crystal type: trigonal
Hardness 3½
Specific gravity: ≈1.9

Cleavage: indistinct
$Ca_3Si(CO_3)(SO_4)(OH)_6 \cdot 12H_2O$

DESCRIPTION: Thaumasite is white or colourless, with vitreous or silky lustre. The crystals are filaments, forming a dense, indistinct, radiating mass. Thaumasite has a very high water content, and low specific gravity. When it loses its water, it almost marks the fingers when touched. Thaumasite is classified with sulphates, and belongs to the so-called ettringite group.

OCCURRENCE: Thaumasite is rare, but has been found with apophyllite, okenite, mesolite and scolecite in olivine basalt. It has been identified in Iceland in samples from Hvalfjörður in the west.

NAME: *Thaumasite*, coined in 1878, is from the Ancient Greek *thaumasio* (= strange).

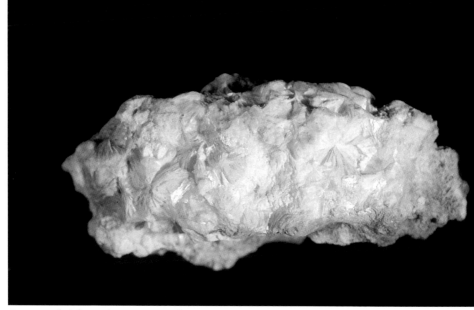

Aggregates of tabular gyrolite, overgrown with reyerite. From *Langavatn, West Iceland*. Size: *1.3 x 3.3 cm*

REYERITE

Crystal type: hexagonal
Hardness: 3½–4½
Specific gravity: 2.54–2.58

Cleavage: perfect in one direction
$(Na,K)_2Ca_{14}(Si,Al)_{24}O_{58}(OH)_8 \cdot 6H_2O$

DESCRIPTION: Reyerite is white, dull or with vitreous or slight pearly lustre. The crystals are laminar, forming small flakes. It crumbles easily, and is distinguishable from most zeolites by being softer.

OCCURRENCE: Reyerite is rare. It occurs in amygdaloidal basalt in the analcime and mesolite-scolecite zones. It is commonly found in association with analcime, along with gyrolite.

NAME: *Reyerite* was named in 1811 after Austrian geologist E. *Reyer*.

From *Hvalfjörður, West Iceland.* Size: *crystals about 2 mm.*

ILVAITE

Crystal type: orthorhombic
Hardness: 5½–6
Specific gravity: 3.8–4.1

Cleavage: none
$CaFe_3OSi_2O_7(OH)$

DESCRIPTION: Ilvaite is black with a slight metallic lustre. It forms prismatic crystals, striated lengthwise, or diamond-shaped polyhedrons. In Iceland, crystals have been found of 2–4 mm diameter, sparsely spaced on quartz and calcite. Streak is black and fracture conchoidal.

OCCURRENCE: Ilvaite is extremely rare in Iceland. It is only known from two localities, in Hvalfjörður in the west, where it occurs in thoroughly altered tholeiitic basalt.

NAME: *Ilvaite* was coined in 1811, from the Latin name of the island of Elba.

Crystal faces and angles are clearly discernible. From *Hvalfjörður, West Iceland*. Size: 0

Ilvaite on quartz. From *Hvalfjörður, West Iceland*. Size: *crystals about 2 mm.*

133

QUARTZ (SILICA) MINERALS

Quartz is one of the commonest minerals in Iceland. It is the second-commonest mineral on earth, after feldspar. It occurs both as a rock-forming mineral, especially in acid igneous rock, and as amygdules, i.e. precipitates from solution. Quartz occurs in more colour variants than any other mineral. Quartz minerals are known in amorphous, cryptocrystalline and crystalline forms, the last of these include large, beautiful crystals. The hardness is 7, except in the case of amorphous silica (opal), which has hardness of $5^{1/2}$–$6^{1/2}$. Quartz is not soluble in acid, except in hydrofluoric acid. The composition of quartz is pure silica, which is coloured by various impurities. Quartz minerals have different names according to their colour, shape and crystal form. Quartz minerals form as amygdules in a large temperature range. Amorphous opal forms at the lowest temperature (20–50°C), while coarsely-crystalline quartz does not form under 150°C. The varied family of quartz minerals is here classified in three groups:

Quartz and variants (crystalline), i.e. rock crystal, amethyst, citrine and smoky quartz.

Chalcedony and variants (cryptocrystalline), i.e. chalcedony, onyx, agate and jasper, and finally opal and variants (amorphous).

Variants of quartz exist with different crystalline structure, i.e. tridymite and cristobalite. Both occur in Iceland, but are very small in size, forming at the final stage of solidification of volcanic rock. Tridymite forms only in rhyolite. Examples of cristobalite have been found that are visible to the naked eye.

White microcrystalline quartz. From *Þormóðsdalur near Reykjavík*. Size: *9 x 8 cm*.

QUARTZ

Crystal type: trigonal (hexagonal)
Hardness: 7
Specific gravity: 2.65

Cleavage: none
SiO_2

DESCRIPTION: Quartz is generally white, milky-white or grey, and translucent or opaque. It has a vitreous lustre and uneven or conchoidal fracture. Quartz forms hexagonal prismatic crystals which end in slanted faces meeting at a point, i.e. a six-sided pyramid. Crystal form may be indistinct or even indistinguishable, e.g. in igneous rock, where quartz occurs as a primary mineral.

OCCURRENCE: Quartz is one of the main rock-forming minerals in acid igneous rocks, but is very small in size except in granophyre and granite. It is also a com-

mon amygdule in tholeiite and silica-rich igneous rocks, in which case it may be transparent, at least at the apex. It is then classified as rock crystal. In the basaltic formation, it is not to be found until one reaches the lower parts of the mesolite-scolecite zone, and then mainly in tholeiite. It may occur as vein filling at shallow depth in geothermal systems.

NAME: *Quartz* is from German. The name was used by medieval German miners for the commonest form of the mineral, greyish quartz that occurs as a gangue mineral.

135

From *Lón, Southeast Iceland*. Size of crystals: *3–5 cm.*

ROCK CRYSTAL

Crystal type: trigonal (hexagonal)
Hardness: 7
Specific gravity: 2.65

Cleavage: none
SiO_2

DESCRIPTION: Rock crystal is a coarsely crystalline form of quartz. It is colourless and transparent, forming prisms with a regular hexagonal pyramid. The prism is generally whitish or greyish, while the point is transparent. In Iceland rock crystal is normally small with short prisms, but may be as long as 10–20 cm. Prism faces are often horizontally striated.

OCCURRENCE: Rock crystal occurs as amygdules in the lower part of the basalt formation, but the best developed specimens are found in high-temperature geothermal systems, preferably in deeply eroded central volcanoes. It is also found as amygdules at the margins of intrusions.

NAME: *Rock crystal* was known as *crystal* (Latin: *crystallum*) in Roman times, when the mineral was believed to be deep-frozen ice, since it was reported to occur near glaciers.

Clusters of rock crystals are most often short in proportion to their width. From *East Iceland*. Size: *7 x 12 cm.*

"Sceptre" variant of rock crystal.
From *Hvalfjörður, West Iceland.* Size: *crystal 25 mm (Photo SSJo).*

Small crystals on chalcedony.
From the *East Fjords.* Size: *crystals about 0.5 cm.*

Rock crystal. From *Snæfellsnes Peninsula, West Iceland*. Size: 7.5 cm.

From *Borgarfjörður, West Iceland.* Size: *cavity 2 x 3 cm.*

CITRINE

Crystal type: trigonal (hexagonal)
Hardness: 7
Specific gravity: 2.65

Cleavage: none
SiO_2

DESCRIPTION: Citrine is a yellowish variant of quartz. In Iceland it is small in size and translucent. The yellow colour is due to iron hydroxide. It is often unclear whether the crystal is yellow throughout or whether the colour is only superficial.

OCCURRENCE: Citrine is very rare in Iceland. It occurs as amygdules adjacent to intrusions. When heated, amethyst becomes yellow. It is possible that citrine in Iceland owes its yellow colour to the heat from intrusions.

NAME: *Citrine* is named for its lemon-yellow colour.

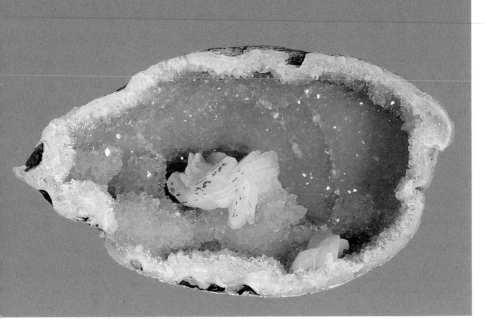

Cavity with amethyst crystals, and calcite at centre. From *Borgarfjörður, East Iceland.* Size: *11.5 x 21 cm.*

AMETHYST

Crystal type: trigonal (hexagonal)
Hardness: 7
Specific gravity: 2.65

Cleavage: none
SiO$_2$

DESCRIPTION: Amethyst is a violet-coloured variant of quartz, which otherwise resembles rock crystal. The colour is believed to be derived from iron. In Iceland amethyst is rather pale in colour. The crystals are transparent or translucent, at least at the apex. The largest amethyst crystals found in Iceland are 15–20 cm in length with a diameter of 4–5 cm at the base, but usually they are much shorter and pro- portionately broader, although the pyramidal form is clear.

OCCURRENCE: Amethyst occurs as amygdules in highly altered igneous rock. It has mainly been found in the east of Iceland, e.g. at Hornafjörður, Lón, Borgarfjörður and Gerpir.

NAME: *Amethyst* is from the Ancient Greek *amethustos* (= sober). Amethysts were used as talismans against drunk- enness.

Amethyst. From *Lón, Southeast Iceland*. Size: *4 x 18 cm*.

From *Lón, Southeast Iceland.* Size of crystals: *2–3 mm.*

SMOKY QUARTZ

Crystal type: trigonal (hexagonal)
Hardness: 7
Specific gravity: 2.65

Cleavage: none
SiO_2

DESCRIPTION: Smoky quartz is a brownish-coloured variant of quartz. In Iceland it is very small in size. It is opaque, and the pyramid is generally darker than the actual crystal prism. The colour is believed to derive from aluminium, or to have been caused by radiation. Rock crystal can be turned into smoky quartz by radiation.

OCCURRENCE: Smoky quartz occurs as amygdules in plutonic rock.

NAME: *Smoky quartz* derives its name from its smoky colour.

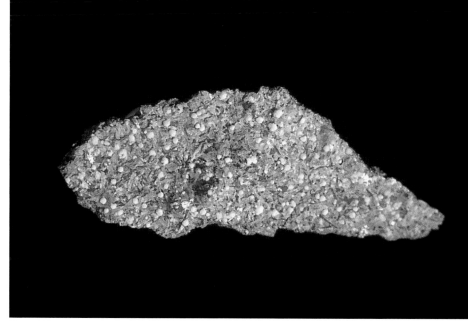

From *Geldinganes near Reykjavík*. Size: *crystals 0.5 x 1 mm.*

CRISTOBALITE

Crystal type: tetragonal, cubic
Hardness: 6½–7
Specific gravity: 2.2

Cleavage: none
SiO_2

DESCRIPTION: Cristobalite is white and without lustre. It forms small individual crystals inside vesicles, and also occurs as coatings.

OCCURRENCE: Cristobalite occurs in vesicles in volcanic rock, formed at the final stage of solidification. It also occurs with opal, both in amygdules and silica sinter, where it can only be identified by X-ray.

NAME: *Cristobalite* is named after *San Cristobal* in Mexico.

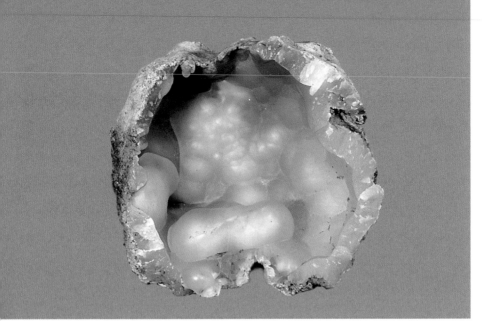

From the *East Fjords*. Size: 6 x 7 cm.

CHALCEDONY

Crystal type: trigonal (hexagonal)
Hardness: 7
Specific gravity: 2.57–2.65

Cleavage: none
SiO_2

DESCRIPTION: Chalcedony is a silica mineral often regarded as amorphous but in fact it forms tiny thread-like crystals which are just distinguishable under a microscope. It is translucent with slight vitreous or greasy lustre. The colour is most commonly whitish or greyish, while other colours have been found, such as pale blue, dark-brown to black, yellowish and greenish. In half-filled cavities chalcedony forms convex crusts, and sometimes knobs several centimetres thick. It is also frequently seen in stalactic form.

OCCURRENCE: Chalcedony is a common amygdule in tholeiitic basalt and rhyolite in Iceland. In cavities it often forms a coating nearest to the rock overgrown by quartz.

NAME: *Chalcedony* is derived from the Greek place-name *Khalkedon*.

Multicoloured chalcedony. The rust colour is due to iron oxide. From the *East Fjords*. Size: *18.5 x 27 cm.*

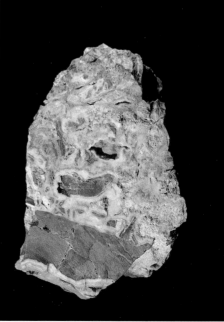

Stalactic chalcedony.
From *Borgarfjörður, West Iceland.* Size: *4 x 7.5 cm.*

Breccia cemented by chalcedony and quartz.
From *Þormóðsdalur near Reykjavík.* Size: *7.5 x 12 cm.*

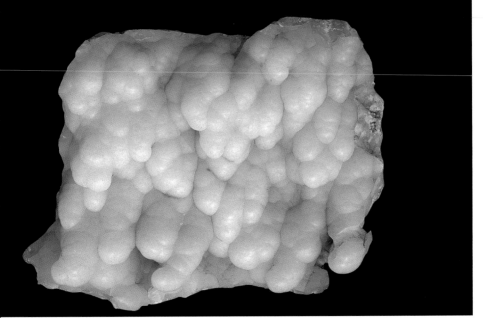

Mamillary chalcedony. From *Southeast Iceland*. Size: *11 x 14.5 cm.*

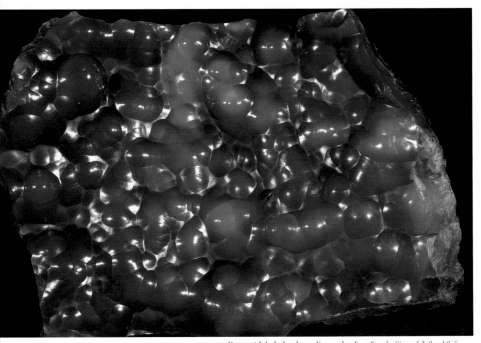

Botryoidal chalcedony. From the *East Fjords*. Size: *13.8 x 18.5 cm.*

From *Southeast Iceland.* Size: *5.2 x 8.4 cm.*

ONYX

Crystal type: cryptocrystalline
Hardness: 7
Specific gravity: 2.65

Cleavage: none
SiO_2

DESCRIPTION: Onyx is a variant of chalcedony with straight parallel bands of different colours. Apart from this it resembles chalcedony, but with a flat, not convex, surface. In Iceland the stripes are often alternately white and colourless.

OCCURRENCE: Onyx is a fairly common amygdule in tholeiitic basalt in Iceland, sometimes occurring with chalcedony and quartz, which then forms the centre of the cavity.

NAME: *Onyx* is from Ancient Greek *onux* (= fingernail).

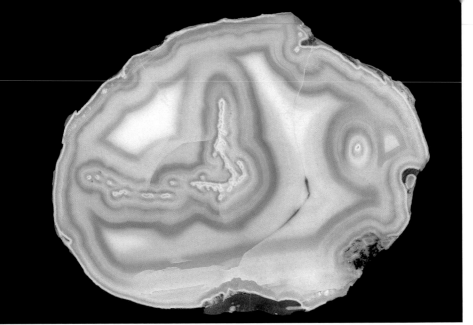

From the *East Fjords*. Size: *11 x 14 cm.*

AGATE

Crystal form: cryptocrystalline
Hardness: 7
Specific gravity: 2.57–2.65

Cleavage: none
SiO$_2$

DESCRIPTION: Agate is a variant of chalcedony with bands of different colours, generally following the contours of the cavity, and hence forming a concentric pattern. Apart from this, agate resembles chalcedony, i.e. as regards lustre, transparency and fracture. In Iceland agate is generally white, grey or pale blue. Irregular patterns also exist, some of which have their own names. One of these is moss agate, where the agate has formed around inclusions of celadonite and/or chlorite.

OCCURRENCE: Agate occurs as amygdules in tholeiitic basalt. It is commonly found with chalcedony and quartz, often in the same cavity.

NAME: *Agate* derives from the name of a river in Sicily.

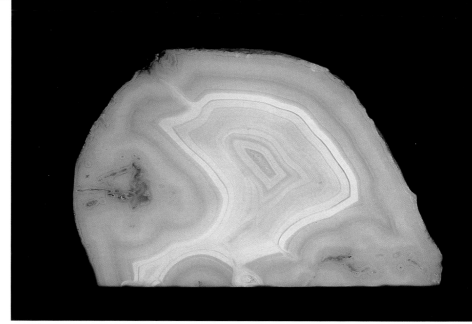

From the *East Fjords*. Size: *12 x 18 cm.*

Moss agate. From *Reyðarfjörður, East Iceland*. Size: *5.6 x 7.7 cm.*

Layered jasper. Quartz lining in the open fissure. From *Lón, Southeast Iceland*. Size: *10 x 14.5 cm.*

JASPER

Crystal type: cryptocrystalline
Hardness: 7
Specific gravity: 2.57–2.65

Cleavage: none
SiO$_2$

DESCRIPTION: Jasper is opaque, without lustre, and always strongly coloured by impurities, generally ferrous compounds and clay. It may be yellow, greenish, red or brown. Several colours may occur in the same sample, irregularly mixed or sometimes forming layers. The crystallinity is just distinguishable under a microscope. It is more coarse-grained than chalcedony, with granular, rather than thread-like, crystals. Jasper forms the largest amygdules that occur in Iceland. Blocks of jasper may weigh 50–100 kg. Petrified wood is usually jasper.

OCCURRENCE: Jasper is a very common amygdule in cavities and fissures in basalt and rhyolite, and is also found as thin layers. A well-known site of this type is at Hestfjall in Borgarfjörður in the west.

NAME: *Jasper* is an ancient name, which can be traced back to Biblical times (Hebrew *yashpeh*).

Multicoloured jasper with chalcedony and onyx.
From the *East Fjords*. Size: *12 x 16 cm.*

Green jasper with red patches. Conchoidal fracture.
From the *East Fjords*. Size: *4.7 x 7.6 cm.*

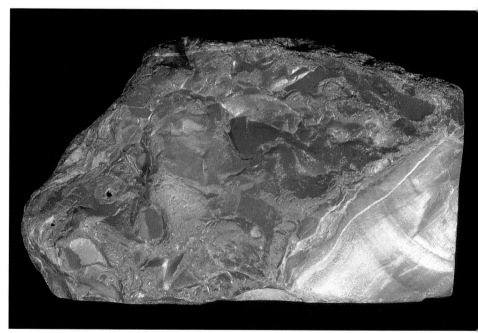

Jasper is most commonly red. From the *East Fjords*. Size: *11 x 18.5 cm.*

151

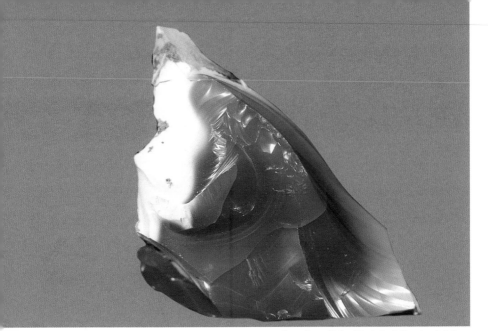

From *Borgarfjörður, West Iceland.* Size: *length 7.2 cm.*

OPAL

Crystal form: amorphous
Hardness: 5½–6½
Specific gravity: 1.9–2.3

Cleavage: indistinct
$SiO_2 \cdot nH_2O$

DESCRIPTION: Opal is a silica mineral that is almost amorphous, with 3–13% water content. Opal is softer and less dense than other silica minerals. The most common variety is pale or milky-white, and opaque or translucent, although transparent opal also occurs. It may be mixed with impurities, in which case it can be grey, brown, greenish or red. It may resemble jasper, but is distinguishable by being softer and more lustrous. It has vitreous or greasy lustre and the fracture is irregular and conchoidal.

OCCURRENCE: Opal is common in Iceland as amygdules in cavities and cracks. It forms at a low temperature and hence is one of the amygdules found in the upper parts of the lava pile, where it occurs in olivine basalt. Opal is a common mineral in active low-temperature geothermal systems. Silica sinter and diatomite are opaline deposits.

NAME: *Opal* is from the Sanskrit *upala* (= stone). Different variants of opal have their own names. Red opal is called *fire opal*. Colourless opal is called *hyalite*. This has been found in Iceland as coatings in vesicles. Opaque, greyish opal, the commonest form, is known as *common opal*. *Precious opal*, which is translucent and multicoloured, is very rare in Iceland.

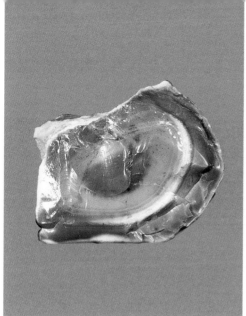

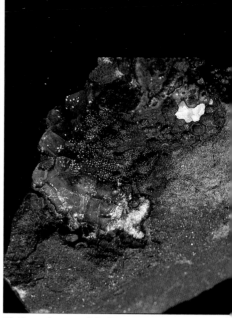

Multicoloured opal. From *Ölfus, South Iceland.*
Size: *6.5 x 5.5 cm.*

Bluish translucent opal (hyalite).
From *Borgarfjörður, West Iceland.* Size: *2 x 3 cm.*

Common opal. From *Mosfellssveit near Reykjavík.* Size: *6 x 7 cm.*

153

CARBONATES, BARYTE AND HALIDES

Carbonates are minerals in which one or more metals form compounds with the acid radical CO_3. About 60 carbonate minerals are known, most of them very rare. About 4% of the earth's crust consists of carbonates, predominantly calcium compounds. Carbonate minerals are highly important to the ecosystem of the oceans, and have an indirect effect upon the quantity of greenhouse gases in the atmosphere. Shells of many organisms consist of calcite and aragonite, as do corals. These organisms take up carbon dioxide from the sea, which can then absorb more carbon dioxide from the atmosphere. The principal carbonates found in Iceland are calcite and aragonite. Both are calcium carbonates, but with differing crystal forms. Calcite and aragonite form the largest crystals in amygdules in Iceland. Siderite, dolomite and breunnerite (a variant of magnesite) have also been found as amygdules in Iceland. Some weathered variants of ore minerals also form compounds with carbonates, e.g. rosasite and malachite, which are here grouped with the ore minerals.

From *Húsafell, West Iceland*. Size: *8.5 x 24 cm.*

CALCITE

Crystal form: trigonal
Hardness: 3
Specific gravity: 2.7

Cleavage: perfect in three directions
$CaCO_3$

DESCRIPTION: Calcite may occur in many different forms. Several hundred crystal forms are known, and thousands of variants. Granular and orthorhombic forms are commonest. Another well-known form is the scalenohedron (dog tooth spar) with triangular facets. Thin lamellar crystals are found in extinct geothermal areas. Calcite is generally white, but may be coloured by impurities. Yellowish, reddish and pink variants are known in Iceland. Calcite normally has vitreous lustre, and it cleaves into rhomboids with 105° and 75° angles. Fracture is conchoidal. Water-containing carbon dioxide dissolves calcite, and occasionally only the imprint of the crystals remains. Calcite effervesces in dilute hydrochloric acid and carbon dioxide gas is released, leaving calcium in the water solution. This is one of the best methods of identifying calcite, along with its cleavage properties.

OCCURRENCE: Calcite is precipitated from water solution at all temperatures. It is one of the commonest amygdules, occurring most frequently around eroded central volcanoes. The largest crystals are found as fissure fillings. At the Mógilsá river on Esja adjacent to Reykjavík, calcite was mined in the late 19th and early 20th centuries. In other countries calcite forms limestone, but true limestone is not known in Iceland.

NAME: *Calcite* is derived from the chemical composition and cleavage of the mineral. The name is from Ancient Greek *chalix* (= chalk).

155

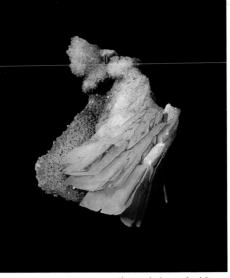

Scalenohedral calcite. From *Grensdalur near Hveragerði,*
South Iceland. Size: *7 x 10 mm (Photo SSJo).*

Flattened, sharp-edged form.
From *Skarðsheiði, West Iceland.* Size: *5 x 7 cm.*

Blocky form with oblique faces. From *Hoffellsdalur, Southeast Iceland.* Size: *36 x 27 cm.*

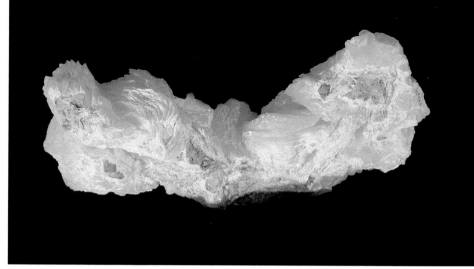

Pink variant. Cluster of coalescing platy crystals. From *Hvalfjörður, West Iceland.* Size: *13 x 10 cm.*

Lenticular twinned crystal on quartz. From the *East Fjords.* Size: *8 x 8 cm.*

Platy variant of calcite. From *North Iceland.* Size: *8.5 x 11 cm.*

From *Hoffellsdalur, Southeast Iceland*. Size: *24 x 21 cm.*

ICELAND SPAR

Crystal type: trigonal
Hardness: 3
Specific gravity: 2.7

Cleavage: perfect, in three directions
CaCO$_3$

DESCRIPTION: Iceland spar is a renowned completely clear variant of calcite. It forms regular crystals, either rhomboids or scalenohedrons (with triangular facets). The crystals are generally matte-surfaced, often with overgrowing zeolites. Iceland spar has vitreous lustre. It cleaves into clear rhomboid crystals with 105° and 75° angles. Fracture is conchoidal. When one looks through the clear crystal, two images are seen, due to the refraction. A Danish physician, Erasmus Bartholin (1625–1698), made extensive studies of refraction using Iceland spar from the Helgustaðir mine in the east. Subsequent research on refraction led to Iceland spar being used in microscopes (Nicol prisms).

OCCURRENCE: The best-known locations for Iceland spar are the Helgustaðir mine at Reyðarfjörður, in the east, which is now a protected site, and Hoffell in Hornafjörður in the southeast.

NAME: For a long time Iceland was the only location where Iceland spar was known to occur, hence the name.

THE LARGEST CRYSTAL OF ICELAND SPAR: The largest known crystal of Iceland spar is exhibited in a glass cabinet in the entrance to the geological collection of the British Museum in London. The crystal, from Helgustaðir, weighs about 4^{1}/$_{2}$ cwt (nearly 230 kg.). It was purchased from a merchant in Reyðarfjörður in 1876. Two circular patches have been polished on each side, and a cross has been painted on one side. When one looks through the crystal, two separate images of the cross are seen, due to the refraction of this unique crystal.

Crystal of Iceland spar at Geological Museum in London. From *Helgustaðir, East Iceland.*
Size: *length about 60 cm (Photo Geological Museum).*

Iceland spar is transparent. It is known for its unique refraction. From *Helgustaðir, East Iceland.* Size: *4.5 x 6 cm.*

From the *East Fjords*. Size: 7.5 x 7.5 cm.

SUGAR CALCITE

Crystal form: trigonal
Hardness: 3
Specific gravity: 2.7

Cleavage: perfect
$CaCO_3$

DESCRIPTION: Sugar calcite is a yellow-ish-brown variant of calcite, which resembles sugar-candy. It also occurs as slender rods. The colour is believed to be due to ferrous compounds. Sugar calcite crystals are commonly 1–4 cm in length.

OCCURRENCE: Sugar calcite is best known in Iceland as infillings of fossil shells of the Tjörnes beds in the north-east. It also occurs in many other locations, as amygdules in basalt.

NAME: *Sugar calcite* is a local Icelandic name which refers to the mineral's resemblance to sugar-candy.

Sugar calcite normally forms rhomboid crystals where it grows freely.
From the *East Fjords*. Size: *5.9 x 5.5 cm.*

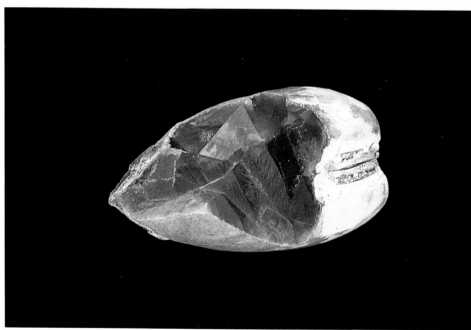

Sugar calcite is common as fillings in shells in the Tjörnes fossiliferous beds.
From *Tjörnes, Northeast Iceland*. Size: *3.3 x 5.6 cm.*

From *Borgarfjörður, West Iceland*. Size: *diameter of largest cluster 5 cm.*

ARAGONITE

Crystal form: orthorhombic
Hardness: 3½–4
Specific gravity: 2.95

Cleavage: distinct
$CaCO_3$

DESCRIPTION: Aragonite has the same composition as calcite, but a different crystal form, which is radiating. The radiating clusters of aragonite are much coarser than those of zeolites, and they often have cracks perpendicular to the long axis. Fracture is conchoidal or jagged. When fresh, aragonite is colourless with vitreous lustre, but it is more generally found in weathered form, when it is greyish with a dull lustre. Aragonite prisms can be up to one cm across, and over 10 cm long.

OCCURRENCE: Aragonite has been found in various locations in Iceland, espe-cially in high-temperature geothermal areas. It also forms as amygdules in igneous rock, especially andesite and basalt. It is unstable at standard state temperature and pressure. If aragonite is heated to 400°C without increased pressure, it recrystallises into calcite. Aragonite occurs in the shells of marine organisms.

NAME: *Aragonite* is named after *Aragon*, a region of Spain.

VARIANTS: *Flos ferri* (= iron flower) is a white variant of aragonite which resembles coral. Its iron content is actually minimal, but it forms due to weathering of ferrous compounds.

Coarse columns of aragonite are commonly found around high-temperature geothermal areas.
From *Heilagsdalsfjall near Mývatn, North Iceland.* Size: *15 x 13 cm.*

Columns of aragonite with clear end faces.
From *Hnappadalur, West Iceland.* Size: *height 4.5 cm.*

Crystal form is not externally visible in the
flos ferri variant of aragonite.
From the *East Fjords.* Size: *7.5 x 7.5 cm.*

From *Lón, Southeast Iceland.* Size: *10 x 6 cm.*

DOLOMITE

Crystal form: trigonal
Hardness: 3½–4
Specific gravity: 2.86

Cleavage: perfect
$CaMg(CO_3)_2$

DESCRIPTION: Dolomite is a calcium-magnesium carbonate similar to calcite, which forms rhomboid crystals, generally of small size. Crystal faces may be curved. Dolomite is clear or white, sometimes reddish, but in Iceland generally yellow due to impurities. It has vitreous or pearly lustre, and is translucent. Fracture is conchoidal and jagged, and streak is white.

OCCURRENCE: Dolomite is very rare in Iceland. It occurs mainly around intrusions, often with ore minerals. In other countries it is common as a mineral occurring on its own, and whole mountains are formed of dolomite (magnesium limestone).

NAME: *Dolomite* was named in 1791 after *D. Dolomieu*, who was the first to distinguish dolomite from calcite.

Quartz coated with dolomite crystals. From *Lón, Southeast Iceland*. Size: *9 x 11.5 cm.*

Pale yellow variant of dolomite in high-temperature environment, with chalcopyrite, rosasite, etc.
From *Össurá, Southeast Iceland*. Size: *crystals about 1 mm.*

From *Borgarfjörður, West Iceland*. Size of spherules: *4 mm.*

SIDERITE

Crystal form: trigonal
Hardness: 4–4½
Specific gravity: 3.7–3.9

Cleavage: perfect
FeCO$_3$

DESCRIPTION: In Iceland siderite occurs most commonly as small spherules, showing a radiating pattern when broken up. It also occurs as tabular crystals, which may be curved. It is yellowish, brown or reddish-brown, but the streak is white or pale brown. Lustre is normally vitreous, but sometimes pearly.

OCCURRENCE: Siderite is fairly rare in Iceland. It is mainly found in fissures in basalt where hot water has flowed, or with various ore minerals at the margins of intrusions. In other countries it is common, and may be an important iron ore.

NAME: *Siderite* is from the Ancient Greek *sideros* (= iron).

Spherical form. Siderite in radiating, partly coalescing spherules on limonite.
From *Borgarfjörður, West Iceland*. Size: *width of chain about 0.5 cm.*

Radiating form. From *Borgarfjörður, West Iceland*.
Size: *diameter of cluster 2 cm.*

Aggregates of small spherules.
From *Borgarfjörður, West Iceland*. Size: 6.5 x 6.5 cm.

Fluorite with platy calcite. From the *East Fjords*. Size: *fluorite crystals 3 x 3 cm.*

FLUORITE

Crystal form: cubic
Hardness: 4
Specific gravity: 3.1–3.2

Cleavage: perfect
CaF_2

DESCRIPTION: Fluorite forms cubes or octahedrons which cleave at right-angles. Twinning is common. Crystals up to 5 cm in size have been found in Iceland. Fluorite also forms spheres and sheaves of laminar radiating crystals. Fluorite is colourless, greenish or pale violet. Many more colour variants occur in other countries, and fluorite is known for this. It is translucent with vitreous lustre and conchoidal fracture. Streak is white. Fluorite is slightly fluorescent, and the phenomenon of fluorescence, first observed in fluorite, is named after the mineral.

OCCURRENCE: Fluorite occurs in a few locations in Iceland, at and near the margins of large granophyre intrusions, and in highly altered rocks in the roots of eroded central volcanoes. It has been found, for instance, at Lýsuhyrna on the Snæfellsnes peninsula. The finest examples are from Breiðdalur in the east.

NAME: *Fluorite*'s name is derived from its chemical composition. Its root is the Latin *fluere* (= to flow), in reference to how easily the mineral melts; it was used in the smelting of steel and aluminium.

Two forms of fluorite with platy calcite. The calcite has formed first, then laminar white fluorite and finally a cube of green fluorite. From *Breiðdalur, East Iceland*. Size of fluorite crystal: *1.4 x 1.6 cm.*

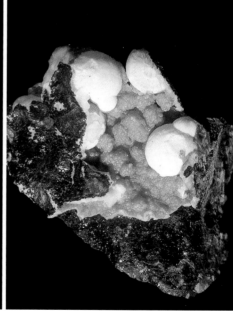

Cubic fluorite crystals with quartz. From *Breiðdalur, East Iceland.* Largest fluorite crystal: *6 x 4.5 cm.*

Spherical clusters of fluorite crystals. From *Hvalfjörður, West Iceland.* Largest sphere: *7 mm.*

169

From Þvottá, Southeast Iceland. Size: 5 mm (Photo SSJo).

BARYTE

Crystal form: orthorhombic
Hardness: 3–3½
Specific gravity: 4.5

Cleavage: perfect
$BaSO_4$

DESCRIPTION: Baryte forms platy, laminar or blocky crystals. It is white or greyish and translucent. Baryte has vitreous lustre and conchoidal fracture. Streak is white. One of its principal features is its weight. Its specific gravity is 4.5, while light-coloured minerals commonly have specific gravity in the range 2.2 to 2.7.

OCCURRENCE: Baryte forms from late magmatic fluids, especially in the vicinity of intrusions. It is very rare in Iceland, but crystals 3–5 cm in size have been found around central volcanoes in three locations in the East Fjords.

NAME: *Baryte* is from the Greek *barus* (= heavy), derived from the mineral's chemical composition (barium) and its weight.

Two forms of baryte. Clustered tabular crystals, on which are large blocky crystals. The brown layer underneath is siderite. From the *East Fjords*. Size: *21 x 27 cm.*

From *Reykjanes, Southwest Iceland*. Size: *6 x 4.5 cm.*

HALITE (Rock salt)

Crystal form: cubic
Hardness: 2
Specific gravity: 2.1–2.2

Cleavage: perfect
NaCl

DESCRIPTION: Halite, or rock salt, is white or grey. It forms cubic crystals, sometimes twinned. In Iceland it is generally small in size, except when precipitated from hydrothermal brines. It dissolves easily in water, and tastes salty, as is generally known. This is its best identifying feature.

OCCURRENCE: Halite is found in many places in nature where sea water or other salt water evaporates. On hyaloclastite formed under the sea, greyish deposits of halite may often be seen, where water seepages from such rock have evaporated. These deposits are washed off by rain, but they form again when the rock dries. Halite occurs in the hyaloclastite mountains in Mýrdalur in the south. It is also known from the Westman Islands, i.e. the island of Surtsey, and Eldfell. The largest crystals occur as precipitates from brine in Reykjanes in the southwest.

NAME: *Halite* is from the Greek *hals* (= salt). The word-component *hall-* occurs in many parts of Europe, at the locations of old salt mines.

ORE MINERALS

While metals are rare in Iceland, various ore minerals occur there. These are sparse, and no exploitable quantities are found. These are generally minerals which are compounds of a metal and oxygen (oxides) or a metal and sulphur (sulphides), in addition to metal carbonates.

Iron minerals are the most common. Magnetite is a rock-forming mineral in igneous rock, which forms other minerals, such as limonite and hematite, upon weathering and alteration. Weathered rock in Iceland is characterised by the reddish-brown colour of iron compounds.

Sulphide minerals are common in extinct geothermal systems and at the margins of intrusions. Weathering of metallic sulphides forms metallic carbonates.

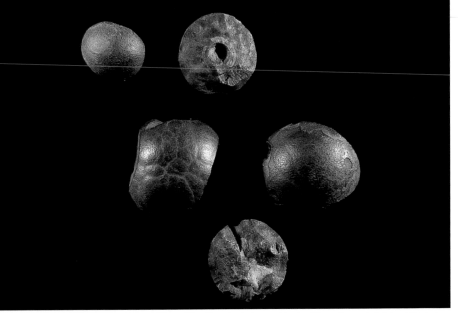

Spherical form. Here limonite is dense with a knobby surface and slight metallic lustre.
From *Skarðsheiði, West Iceland*. Size: *diameter about 1 cm.*

LIMONITE

Crystal form: near-amorphous
Hardness: 4–5½
Specific gravity: 2.7–4.3

Cleavage: indistinct
FeOOH·nH$_2$O

DESCRIPTION: Limonite is dark or yellowish brown, sometimes reddish, opaque, without lustre or with a slight metallic lustre. It may be dense, forming thin layers and coatings with a globular surface, or earthy and friable. In the dense, globular variant radiating crystals may be discernible. Limonite is a mixture of natural iron oxides, but not a true mineral. The iron oxides in limonite are predominantly goetheite and its variants.

OCCURRENCE: Limonite forms by the oxidation of ferrous minerals, such as magnetite, which is found in all basalts.

NAME: *Limonite* is from the Latin *limus* (= marsh), with reference to bog iron. *Goethite* derives its name from the great German writer Goethe, who was interested in geology and mineralogy.

VARIANTS: *Bog iron* is a variant of limonite. It forms when acids in the soil dissolve iron from rock. It is then carried by water, and precipitates under oxidising conditions. The iron content of bog iron may be as high as 65%. In olden times it was used as a source of iron. *Rust* has the same composition as limonite. *Yellow ochre* is a powdery yellow or brown variant of limonite.

174

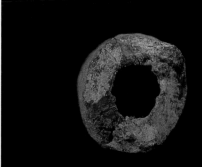

Bog iron coating on calcite.
From *East Iceland*. Size: *6 x 9.5 cm.*

Deposit of limonite from marsh water around a stalk.
From *Krýsuvík, Southwest Iceland*. Size: *diameter 1 cm.*

Limonite. Impregnation in fine-grained sediment. From *Snæfellsnes Peninsula, West Iceland*. Size: *7.8 x 8.2 cm.*

Deposit of limonite in rhyolitic tuff.
From *Óshlíð, West Fjords*. Size: *8 x 4 cm.*

Limonite with spherules of concentric layers.
Radiating form is just discernible. From the
East Fjords. Size: *10.3 x 5.3 cm.*

175

Knobby variant of hematite. From *Hvalfjörður, West Iceland*. Size: *3.5 x 4.5 cm.*

HEMATITE

Crystal form: trigonal
Hardness: 5–6
Specific gravity: 5.3

Cleavage: none
Fe_2O_3

DESCRIPTION: Hematite is an iron oxide devoid of water. It occurs in various forms, some of which have their own names. When coarsely crystalline it is steel-grey or blackish, sometimes with a bluish or multicoloured sheen. When cryptocrystalline it is red or reddish-brown. When coarsely crystalline it has a metallic lustre, but is otherwise without lustre, matte and opaque. Streak is reddish-brown. In Iceland hematite is rarely found in clearly crystalline form, greyish and shiny. In this form it is best known on stalactites in lava caves. The cryptocrystalline, reddish variant is very common as a groundmass between grains in the red interbeds of the basalt formation.

When heated it becomes magnetised. The red interbeds are thus strongly magnetic at the contact surface with the lava layers that cover them.

OCCURRENCE: Hematite is formed by oxidation of magnetite in igneous rock, as deposits around hot springs, and as sublimates from volcanic exhalations. In other countries, hematite is the principal iron ore.

NAME: *Hematite* is from the Greek *haimatites* (= clotted blood), with reference to its blood-like colour.

VARIANTS: *Red hematite* is a microcrystalline variant of hematite which is characteristic of the red interbeds. *Red ochre* is a powdery, reddish variant of hematite, often mixed with clay.

176

Hematite with multicoloured sheen. From *Drápuhlíðarfjall, Snæfellsnes peninsula, West Iceland.* Size: *5 x 3.5 cm.*

Steel-grey hematite on the outside and inside of stalactites. From *Svínahraun, Southwest Iceland.* Size: *thickest stalactite 5 cm in length.*

Well-developed crystals of tabular hematite. From *Hvalfjörður, West Iceland.* Size: *cluster 12–15 mm (Photo SSJo).*

Large magnetite crystals in olivine basalt (basalt-pegmatite). From *Geldinganes near Reykjavík*. Size: 7.7 x 6 cm.

MAGNETITE

Crystal form: cubic
Hardness: 5½–6
Specific gravity: 5.2

Cleavage: indistinct
Fe_3O_4

DESCRIPTION: Magnetite forms small black cubic crystals or octahedrons. One of its main identifying features is magnetism. It is opaque with metallic lustre, and streak is black. In Iceland magnetite is generally very small in size. The largest examples are found in basalt-pegmatite, i.e. coarse-grained veins and vesicular trains in lava layers.

OCCURRENCE: Magnetite is a primary mineral in igneous rocks. It is most abundant in basalt and andesite. The magnetism of these rocks is due to magnetite. It also occurs in veins in geothermal areas, and where magma residual with high gas content has forced its way into voids in solidifying lava.

NAME: *Magnetite* is named for its magnetism. Originally the word is derived from *Magnesia* in Macedonia.

Cluster of skeletal magnetite which has crystallized from magma residue in a lava cave. From *Raufarhólshellir, South Iceland*. Size: *largest crystal 2 mm (Photo SSJo)*.

Magnetite mixed with maghemite (magnetic hematite) of sedimentary origin, baked by molten basalt. From *Borgarfjörður, West Iceland*. Size: *5 x 7 cm*.

From *Gljúfurá, North Iceland.* Size: *1.5 x 1.5 cm.*

PYRITE

Crystal form: cubic
Hardness: 6–6½
Specific gravity: 4.9–5.1

Cleavage: none
FeS$_2$

DESCRIPTION: Pyrite generally forms yellowish cubic crystals which are usually small in size. However, crystals measuring 3–5 cm across have been found in Iceland. Dodecahedral crystals may also form, but these are very rare. Streak is black. Pyrite has a strong metallic lustre on fresh crystal surfaces. Weathering produces a multi-coloured sheen.

OCCURRENCE: Pyrite is common in Iceland. It is most abundant in highly altered rock of extinct central volcanoes, where it forms shiny coatings on the surface of fissures, or clusters of glittering golden granules in the rock. It also occurs in dikes, and is common

at geothermal springs, especially in high-temperature areas. A grey film which often floats on geothermal mud-pots consists of small pyrite crystals.

NAME: *Pyrite* is from the Greek *pyr* (= fire), relating to the fact that it gives off sparks when ground. It is also known as *fool's gold* for its superficial resemblance to gold.

VARIANTS: *Marcasite* is a rhombic iron sulphide. It forms at lower temperatures than pyrite, and occurs with it. It is principally distinguished from it by the streak, which is dark green. The name *marcasite*, from the Arabic, means *firestone*.

Cubic form makes for stepped faces in large crystal. From *Gljúfurá, North Iceland*. Size: *2.5 x 2.5 cm*.

Pyrite crystals on platy calcite. Size: *10 x 10.5 cm*.

Small pyrite in altered palagonite tuff. From *Ölfus, South Iceland*. Size: *height 6 cm*.

Pyrite forming multiface crystals. From *Hvalfjörður, West Iceland*. Size: *crystals about 1 mm*.

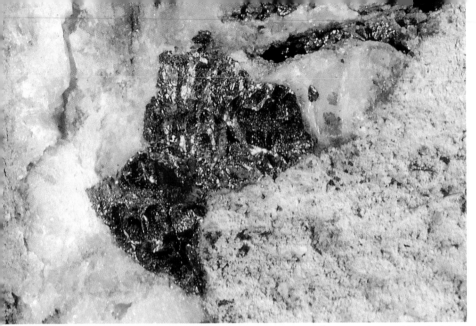

Chalcopyrite without clear crystal form. From *Össurá, Southeast Iceland*. Size: *1 x 1.2 cm*.

CHALCOPYRITE

Crystal form: tetragonal
Hardness: 3½–4
Specific gravity: 4.2

Cleavage: indistinct
$CuFeS_2$

DESCRIPTION: Chalcopyrite is yellow in colour, forming octahedrons, which are, however, usually greatly distorted. In Iceland it is often found as massive aggregates. It is distinguished from pyrite by darker colour and redder weathered surfaces, and it is also not as hard. Streak is slightly greenish black.

OCCURRENCE: Chalcopyrite forms from late magmatic gas-rich liquids. It occurs mainly in veins close to the margins of large intrusions. The best-known location is at the Össurá river in Lón in the southeast.

NAME: *Chalcopyrite* is from the Greek *khalkos* (= copper) and *pyr* (= fire).

Chalcopyrite with clear crystal faces. From *Lón, Southeast Iceland.* Size: *width 2.5 cm.*

Multicoloured chalcopyrite filling interstices in breccia. From *Lón, Southeast Iceland.* Size: *8.5 x 6 cm.*

From *Össurá, Southeast Iceland. Size: largest crystal 1.2 x 1 cm.*

SPHALERITE

Crystal form: cubic
Hardness: 3½–4
Specific gravity: 3.9–4.1

Cleavage: perfect
ZnS

DESCRIPTION: Sphalerite is the commonest zinc mineral. It forms yellowish or brown translucent flakes or larger opaque crystals, sometimes distorted octahedrons. It may contain traces of iron, in which case it is darker in colour. Sphalerite has greasy or metallic lustre. Streak is pale brown or white. In Iceland sphalerite crystals are commonly 0.5–1 cm in size, but larger examples have been found. It is distinguished from chalcopyrite by its cleavage properties and from galena by its colour.

OCCURRENCE: Sphalerite is formed from late magmatic gas-rich liquids, and is mainly found in veins near the margins of large intrusions, like most other metal sulphide minerals.

NAME: *Sphalerite* is from the Greek *sphaleros* (= deceptive), because it is easily mistaken for other minerals.

From *Lón, Southeast Iceland*. Size: *crystal cluster of galena c. 2 cm in diameter.*

GALENA

Crystal form: cubic
Hardness 2½–3
Specific gravity: 7.5

Cleavage: perfect
PbS

DESCRIPTION: Galena is the commonest lead mineral. It forms leaden-grey, shiny cubic crystals, usually rather small. Galena is opaque with strong metallic lustre, and streak is leaden-grey, almost black. It is one of the heaviest minerals found in Iceland.

OCCURRENCE: Galena is formed from late magmatic gas-rich liquids, and is found mainly in veins at the margins of large intrusions, like other metal sulphide minerals.

NAME: *Galena* is Latin for *lead ore*.

From *Krýsuvík, Southwest Iceland*. Size: *3.3 x 4.2 cm.*

COVELLITE

Crystal form: hexagonal
Hardness: 1½ –2
Specific gravity: 4.7

Cleavage: perfect
CuS

DESCRIPTION: Covellite forms thin tabular crystals with metallic lustre. These are commonly blue or purple, but may also be pale or dark blue. Streak is dark blue.

OCCURRENCE: Covellite is a common copper-sulphide mineral which forms by weathering of other copper sulphides, and is found mainly with chalcopyrite adjacent to the large intrusions in the southeast. It also occurs as sublimates from magmatic gas exhalations, and as a coating near geothermal steam vents.

NAME: *Covellite* is named after Italian geologist *N. Covelli*, who first identified it in the early 19th century.

From *Össurá, Southeast Iceland*. Size: *crystal c. 1 cm.*

MALACHITE

Crystal form: monoclinic
Hardness: 3½–4
Specific gravity: 4

Cleavage: distinct
$Cu_2(OH)_2CO_3$

DESCRIPTION: Malachite is a copper-hydroxide carbonate. It forms bright green crusts or coatings on chalcopyrite, which can easily be scraped off with a knife. Sometimes it forms small aggregates which may be just over half a centimetre in size. The inner structure is fibrous and layered. In Iceland crystals are hardly distinguishable in malachite. It is opaque with slight vitreous lustre, and streak is pale green.

OCCURRENCE: Malachite occurs as a weathering product of copper minerals, and is found with chalcopyrite.

NAME: *Malachite* is from *malache*, the Greek name of the mallow plant.

From *Össurá, Southeast Iceland*. Size: *spherules 1–2 mm (Photo SSJo)*.

ROSASITE

Crystal form: monoclinic
Hardness: 4–4½
Specific gravity: 4–4.2

Cleavage: perfect in one direction,
but indistinct
$(Cu,Zn)_2CO_3(OH)_2$

DESCRIPTION: Rosasite is a copper-zinc carbonate mineral formed by weathering of copper and zinc minerals. It is blue-green in colour, like many other secondary copper minerals. Rosasite occurs as coatings and spherules, with fibrous texture visible on fracture surfaces. It is translucent with vitreous lustre, or matte. Streak is pale green.

OCCURRENCE: Rosasite occurs with chalcopyrite and sphalerite, mainly at the margins of large intrusions.

NAME: *Rosasite* is from the name of the *Rosas* mine in Sardinia where the mineral was first found.

When rock is buried by a thick pile of strata, it heats up and is altered. In Iceland erosion of the rock pile extends only 1,500 to 2,000 metres below the original surface, so Icelandic rock exhibits only the first stage of alteration, which takes place under relatively low pressure (up to 4 Kbars) and up to 300–400°C. This produces low-temperature alteration zones, which are typified by zeolites, and below them high-temperature zones with other characteristic alteration minerals, so-called high-temperature minerals, while the zeolites vanish. The greatest alteration is seen in extinct and eroded high-temperature areas, where the rock is characteristically greenish. Such rock was formerly known as propylite. The colour is mainly due to chlorite and epidote, which are typical for alteration at 200–300°C. Other minerals that may form at 300°C or over are prehnite, garnet, actinolite, wollastonite and hedenbergite. Wollastonite is a common mineral occurring with garnet and epidote in high-temperature systems, but in Iceland it has hitherto only been found in borehole cuttings. It is small-scale, fibrous and white, and it may be just discernible with a magnifying glass.

From *Southeast Iceland*. Size: *width 5.5 cm.*

EPIDOTE

Crystal form: monoclinic
Hardness: 6–7
Specific gravity: 3.3–3.5

Cleavage: perfect in one direction
$Ca_2(Fe,Al)Al_2Si_3O_{12}(OH)$

DESCRIPTION: Epidote is generally micro-crystalline, forming a green or yellowish-green coating on the walls of cavities and fissures. Streak is white or slightly grey. In half-filled cavities it may form small elongated prismatic crystals, 0.2–0.5 cm in length, with a bluntly pointed end. It may have discernible vitreous lustre, it is translucent or opaque, and fracture is irregular or conchoidal.

OCCURRENCE: Epidote forms in both basaltic and acidic rocks at temperatures around 230°C and above. It is thus mainly found in eroded, extinct volcanoes, which hosted high-temperature geothermal systems while active. NAME: *Epidote* was coined in 1801 from Greek *epidosis* (= addition), with reference to how the crystals grow.

Epidote needles. From *Hafnarfjall in Borgarfjörður, West Iceland*. Size: *length of crystals 0.5–1 mm (Photo SSJo).*

Crusts of small crystals of epidote. From *Lón, Southeast Iceland.* Size: *5.5 x 7.5 cm.*

From *Lón, Southeast Iceland*. Size: *knobs 2.5 mm.*

PREHNITE

Crystal form: orthorhombic
Hardness: 6–6½
Specific gravity: 2.9

Cleavage: distinct
$Ca_2Al_2Si_3O_{10}(OH)_2$

DESCRIPTION: Prehnite forms small spherical clusters of crystals with vitreous lustre, often pale green in colour, but sometimes white or greyish. The crystal clusters are radiating, and this is visible when broken up. The fracture is irregular and streak is white.

OCCURRENCE: Prehnite occurs with epidote and other high-temperature minerals in basaltic rock in the roots of deeply eroded central volcanoes. It forms at 250°C and higher temperatures, and hence it is found at deep levels in the high-temperature alteration zone.

NAME: *Prehnite* was named in 1774 after *van Prehn*, a Dutchman who brought this mineral from Africa.

Prehnite as an amygdule in gabbro. From *Lón, Southeast Iceland.* Size: *knobs 2–5 mm.*

From *Ferstikla in Hvalfjörður, West Iceland*. Size: *length 4.5 cm.*

GARNET

Crystal form: cubic
Hardness: 6½–7½
Specific gravity: 3.4–3.6

Cleavage: indistinct
$Ca_3Al_2[SiO_4]_3$–$Ca_3Fe_2[SiO_4]_3$

DESCRIPTION: Garnet is a group of minerals of varying composition. In Iceland garnet is mainly andradite (iron-rich), grossular (calcium-rich) and hydrogrossular, the only form of garnet that contains water. Garnet forms small, regularly shaped multi-faced crystals with vitreous lustre. They grow individually, though densely packed together, and are easily identified by the crystal form. Colour is generally reddish-brown or brown. Streak is white. Fracture is conchoidal. Garnet is the hardest mineral found in Iceland.

OCCURRENCE: Garnet occurs mainly in highly altered rock adjacent to intrusion complexes. It has been found in eroded extinct geothermal areas, in the roots of central volcanoes, often with epidote, but principally in fissures close to intrusions. It forms only at 300°C and above. It has been found, for instance, at the Setberg volcano on the Snæfellsnes peninsula, the Hvalfjörður volcano in the west and in the southeast in the alteration halo around large intrusions, e.g. Geitafell, Vestrahorn and Slaufrudalur.

NAME: *Garnet* derives its name from its resemblance to the seeds of the pomegranate (Latin *granatum* = pomegranate).

Well-developed crystal form with striated faces. From *Ferstikla in Hvalfjörður, West Iceland*.
Size: *crystals about 1 mm (Photo SSJo)*.

Unusually large garnet crystals. From *Lón, Southeast Iceland*.
Size: *diameter of crystals about 6 mm*.

Fibrous actinolite with hackly fracture surfaces. From *Hafragil in Lón, Southeast Iceland*.
Size: *crystal at centre right of photo 0.5 cm.*

ACTINOLITE

Crystal form: monoclinic
Hardness: 5½–6
Specific gravity: 2.9–3.3

Cleavage: perfect
$Ca_2(Mg,Fe)_5Si_8O_{22}(OH,F)_2$

DESCRIPTION: Actinolite belongs to the amphibole group. Crystals are generally very slender fibres, often radiating with vitreous lustre. Actinolite is usually greenish, white or grey, and streak is white. Fracture is irregular or jagged. The largest examples, found in cavities in gabbro, are 0.5–1 cm long.

OCCURRENCE: Actinolite forms at temperatures of 300°C or more, in and adjacent to intrusions or deep in high-temperature geothermal areas. It often appears to grow out of pyroxene in altered gabbroic rock, e.g. at Hafragil, Lón, in the southeast.

NAME: *Actinolite* is from the Greek *aktis* (= ray) and *lithos* (= stone).

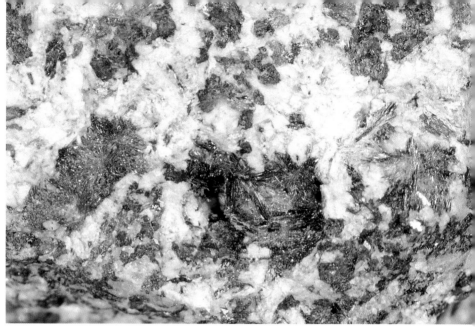

Gabbro with actinolite fibres. From *Hafragil in Lón, Southeast Iceland*. Size: *dark crystals 2–4 mm.*

Greenish actinolite and chlorite retain the form of pyroxene crystals in altered gabbro. Plagioclase (light in colour) is largely altered into albite. From *Vestrahorn, Southeast Iceland*. Size: *largest crystal 7.8 x 12 mm..*

197

From *Geitafell, Hornafjörður, Southeast Iceland*. Size: 7.5 x 5.5 cm.

HEDENBERGITE

Crystal form: monoclinic
Hardness: 5½–6
Specific gravity: 3.2–3.6

Cleavage: perfect in two directions
$Ca(Fe,Mg)Si_2O_6$

DESCRIPTION: Hedenbergite is a variant of pyroxene which forms as slender dark-green rods with vitreous lustre. Streak is white to pale green. Crystals up to 0.5 cm in length have been found in Iceland.

OCCURRENCE: Hedenbergite forms as amygdules at high temperatures, i.e. 400°C and above. In Iceland it has been found near the contact of host rock with intrusions in the southeast.

NAME: *Hedenbergite* was named after the Swedish chemist who first analysed it.

CLAY MINERALS

The term *clay* is used both of a group of minerals, and of very fine-grained rock particles. All clay minerals contain hydroxide or water. The chemical composition may vary in the same mineral. They are so-called sheet silicates, with loosely bound hydroxide between the layers. Crystals are laminar, with perfect cleavage, but are very small-scale and can only be distinguished under a microscope or by X-ray. Clay minerals absorb water in a damp environment, and water is easily released when they dry out. This means that their specific gravity is variable. Clay minerals have hardness of 1–3, and most can be marked by a fingernail. In Iceland clay is mainly formed by alteration of rock in geothermal areas, both on the surface and deep in the rock strata. It is also formed by weathering, especially in warm humid conditions such as those that prevailed in Iceland in the Miocene epoch, during formation of the oldest basalts of the Northwest Peninsula. Where steam is emitted in geothermal areas, the rock around the vents breaks down completely. Such geothermal clay is sticky, soft and generally bluish or grey. In Iceland geothermal clay has been used in pottery. Clay plays an important role in soil, as it absorbs water.

From *Krýsuvík, Southwest Iceland*. Size: 5 x 3 cm.

KAOLINITE

Crystal form: triclinic
Hardness: 2–2½
Specific gravity: 2.6

Cleavage: perfect
$Al_2Si_2O_5(OH)_4$

DESCRIPTION: Crystal form cannot be distinguished in kaolinite with the naked eye or a magnifying glass. The crystals are minute and laminar. Kaolinite forms whitish layers and masses where e.g. geothermal water and steam have altered feldspar-rich rock. This is known as China clay. It is dense, supple and sticky when damp, and it can easily be kneaded and moulded in the hand. In Iceland China clay is generally contaminated with impurities, especially iron oxides, in which case it has brown streaks or spots.

OCCURRENCE: Kaolinite is mainly found at steam vents in high-temperature areas. It is found in its purest form around acidic hot springs in rhyolitic lava, e.g. Hrafntinnusker in the south central highlands. A layer of kaolinite has been found in Mókollsdalur in Strandasýsla in the northwest.

NAME: *Kaolinite* is from the Chinese *Kauling* (= high ridge), a place near Jauchau Fu in China, which was an ancient source of China clay for porcelain production.

Altered basalt. Smectite comprises more than 70% of the specimen. From *Ölfus, South Iceland*. Size: *6 x 8.5 cm*.

Smectite crust on basalt core. From *Ölfus, South Iceland*. Size: *3.8 x 2.5 cm*.

SMECTITE (Montmorillonite)

Crystal form: monoclinic
Hardness: 1–2
Specific gravity: 2.5 (variable according to water content)

Cleavage: perfect
$(Na,Ca)(Al,Mg)_6(Si_4O_{10})_3(OH)_6 \cdot nH_2O$

DESCRIPTION: Smectite (montmorillonite) is the name of a group of clay minerals of variable water content. They easily absorb water, and expand considerably. Smectite is generally brownish or greenish, but a black variety also occurs. It is found in patches or as thin coatings in altered rock, and indistinct crystallisation may be seen under a microscope.

OCCURRENCE: Smectite forms in geothermal areas, both underground and at the surface, through alteration of all types of rock in an alkaline environment. It is found at alkaline hot springs. In Iceland smectite is the most abundant secondary mineral. It forms by alteration of olivine, pyroxene, calcium-rich feldspar and glass, from low temperatures up to 170°C. It plays an important role in soil, as it absorbs water.

NAME: *Montmorillonite* was named in 1847 after the place where it was found, *Montmorillon* in France. *Smectite* is an older name (from 1788), from the Greek *smektes* (= to rub). It was found later that both were the same mineral. Smectite is also known as *fuller's earth*, which was used in the process of washing (fulling) wool, as it absorbs oils.

201

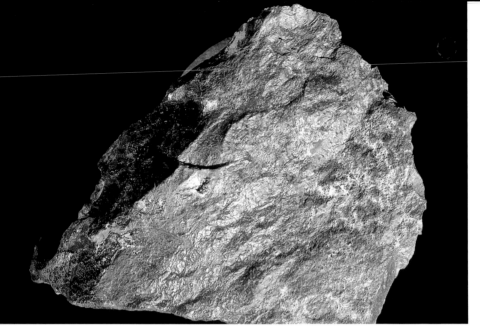

From *Hvalfjörður, West Iceland.* Size: *8 x 5 cm.*

CHLORITE

Crystal form: monoclinic
Hardness: 2–3
Specific gravity: around 2.8

Cleavage: perfect
$(Fe,Mg,Al)_6(Si,Al)_4O_{10}(OH)_8$

DESCRIPTION: Chlorite is classified here with clay minerals, as it is formed similarly to them. It is hardly distinguishable from smectite except by X-ray. Chlorite is greenish to brown, and streak is green. It forms coatings and lumps, often flaky, with flexible (not plastic) flakes. Internal structure is fibrous.

OCCURRENCE: Chlorite forms in considerable quantity at some depth in geothermal areas where temperatures have exceeded 200°C. At lower temperatures it forms together with smectite. It is common, formed by alteration of rock-forming minerals, and as amygdules. It also occurs on fissure surfaces. A green colour characteristic of eroded central volcanoes is generally attributable to chlorite.

NAME: *Chlorite* is from the Greek *chloros* (= green).

VARIANTS: *Chlorophaeite* is similar to chlorite in composition (hydrated Ca, Mg, Fe silicate). It is dark green, and in weathered basalt it is brown or black. Chlorophaeite is waxy with greasy lustre, and it can be marked with a fingernail. It is common in amygdules in tholeiite, as coatings or patches in filled or half-filled cavities.

Basalt largely altered into chlorite. From *Hafnarfjall in Borgarfjörður, West Iceland*. Size: *3 x 2.5 cm.*

Chlorophaeite. Amygdules shrink and crack as they dry. From *Hvalfjörður, West Iceland*. Size: *largest cavity 1.5 cm.*

Chlorophaeite. From *Hvalfjörður, West Iceland.*

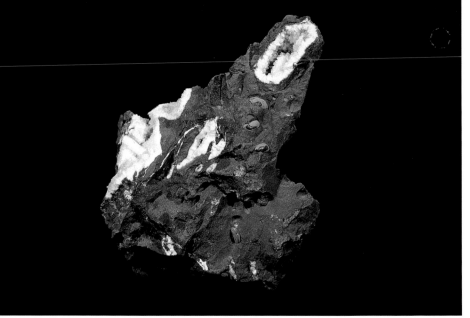

From *Teigarhorn, East Iceland*. Size: *13.5 x 18 cm.*

CELADONITE (Illite)

Crystal form: monoclinic
Hardness: 2
Specific gravity: 2.95–3.05

Cleavage: perfect
$K(Mg,Fe)(Fe,Al)[(OH)_2 | Si_4O_{10}]$

DESCRIPTION: Celadonite is most closely related to a group of minerals known as illite, an intermediate stage between smectite and mica. Celadonite is clayey and blue-green in colour. Crystal form, as in other clay minerals, cannot be discerned with the naked eye or a magnifying glass.

OCCURRENCE: Celadonite is common in Iceland. It occurs as a thin greenish coating on amygdules. Green spots on chalcedony, and green coatings on zeo-lites nearest to the cavity wall, are generally celadonite. True illite occurs in Iceland with smectite in interbeds of the basalt formation particularly in acid tuff and ignimbrite. Illite is the principal mineral in brick clay. In Iceland it is hardly pure enough for brick-making.

NAME: *Celadonite* was coined in 1847 from the French *celadon* (= sea-green). *Illite* was coined in 1937, from the name of the state of *Illinois* in the USA.

In geothermal areas, a large variety of minerals form when surface rock is altered or dissolved, and new minerals are precipitated. Some of these minerals have been mentioned elsewhere in this book, such as pyrite, covellite and the clay minerals kaolinite and smectite. Variants of other common minerals have also been mentioned, such as calcite and aragonite in the form of travertine or tufa, opal in the form of silica sinter/geyserite and hematite deposited around hot springs. In addition to these, there are several common minerals which form under these conditions. These are described in this section, together with the variants of minerals that are typical for hot springs.

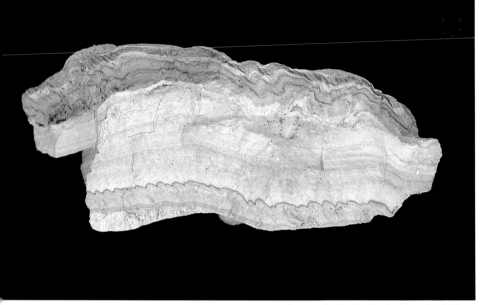

From *Haukadalur, South Iceland*. Size: *length 12 cm.*

SILICA SINTER, GEYSERITE (Opal)

Crystal form: amorphous
Hardness: 5½–6½
Specific gravity: 1,9–2,3

Cleavage: indistinct
$SiO_2 \cdot nH_2O$

DESCRIPTION: Geothermal water contains more silica in solution (SiO_2) than cold water. The higher the temperature of the geothermal system, the higher the silica content. The silica precipitates out of solution when the water cools as it reaches the surface. Silica sinter occurs only at hot-water springs where the temperature of the system is well over 100°C. Its composition is that of opal. Sometimes the silica is deposited on vegetation, leaving marks of leaves, grass and twigs.

OCCURRENCE: Silica sinter forms principally around hot springs in high-temperature areas, e.g. at Hveravellir in the central highlands, and at Geysir and Hveragerði. It also occurs in far smaller quantities in low-temperature areas, e.g. around Laugar in Hrunamannahreppur in the south, at Leirá in the west (mixed with travertine), and in Dufansdalur, Arnarfjörður in the West Fjords.

NAME: *Geyserite* derives its name from the Great Geysir.

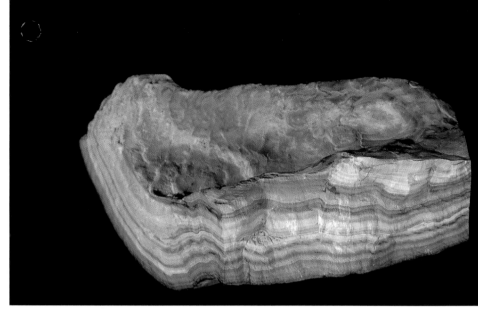

From *Hengill, Southwest Iceland*. Size: *6 x 13 cm*.

TRAVERTINE, TUFA (Calcite)

Crystal form: trigonal
Hardness: 3
Specific gravity: 2.7

Cleavage: perfect in three directions
$CaCO_3$

DESCRIPTION: Water in geothermal systems generally contains carbon dioxide, which dissolves calcium, iron and many other substances from the rock. The calcium may then precipitate out at the surface as calcium carbonate, when the geothermal water boils, and when it evaporates. This forms a crust with distinct layering around hot springs which is called travertine or tufa. Some mineral springs are effervescent with carbon dioxide. Very often their water is contaminated with iron. Travertine is most commonly found around such hot mineral springs, often discoloured with iron precipitates.

OCCURRENCE: Travertine is deposited around carbon-dioxide-rich hot springs. The best examples are in Laugarvalladalur in the east, Blautukvíslabotnar in the central highlands, and at the Lýsuhóll mineral spring and at Hraunsfjörður on the Snæfellsnes peninsula. Some plant remains are commonly found in travertine.

NAME: *Travertine* is a modification of the Italian *tivertino*, from the old Roman name of Tivoli, where travertine forms an extensive deposit.

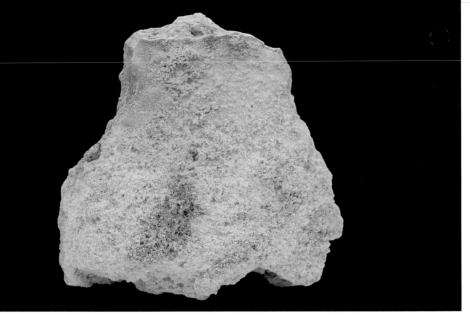

From *Southwest Iceland*. Size: *11 x 12 cm.*

SULPHUR

Crystal form: orthorhombic
Hardness: 2
Specific gravity: 2

Cleavage: indistinct
S

DESCRIPTION: Sulphur forms a dense yellow mass, often with discernible rod-shaped crystals with a pyramidal point, sometimes also with granular shape. The colour is bright yellow, but dense masses may be pale yellow. Due to clay and impurities in geothermal areas, it may be reddish, green, brown or grey. Streak is yellow. Sulphur is often soft and friable, and must therefore be handled carefully when sampled. In a humid environment it forms hydrogen sulphide, which produces the familiar "rotten egg" odour of hot springs. Sulphur crumbles at room temperature. It burns and melts at just over 100°C.

OCCURRENCE: Sulphur forms at fumaroles in high-temperature geothermal areas. Sulphur has been used in the production of gunpowder, explosives and matches, and in rubber manufacture. Sulphur was mined and exported from Iceland in the 13th to 18th centuries. Initially the archbishop of Nidaros (modern Trondheim) had a monopoly on Iceland's sulphur exports. Since gunpowder was not invented until about 1400, it is not known what sulphur was used for before that time. The church is reputed to have transported sulphur all over Europe, and burned it to demonstrate what Hell would smell like.

NAME: *Sulphur* is from the Latin *sulfur*. An obsolete name for sulphur is *brimstone (burn + stone)*.

Granular sulphur. From *Krýsuvík, Southwest Iceland*. Size: *7 x 8.7 cm.*

From *Krýsuvík, Southwest Iceland*. Size: *9.3 x 6.7 cm.*

GYPSUM

Crystal form: monoclinic
Hardness: 2
Specific gravity: 2.3–2.4

Cleavage: perfect in one direction
$CaSO_4 \cdot 2H_2O$

DESCRIPTION: Gypsum is white, colourless or greyish, forming prismatic or tabular crystals. They are commonly intergrown to form crusts. Lustre is vitreous or pearly. Gypsum is easily marked with a fingernail. Fracture is irregular if distinguishable.

OCCURRENCE: Gypsum forms in high-temperature areas adjacent to hot springs and fumaroles where sulphur-rich gases are emitted forming compounds with calcium from the rock.

NAME: *Gypsum* is from the Greek *gypsos* (= chalk).

VARIANTS: *Anhydrite* has the same composition as gypsum, but is without water. It is white or greyish, forming flakes with vitreous or pearly lustre. Hardness is 3–3½. Anhydrite forms at some depth in high-temperature areas, at 200°C and over. It may also form at lower temperatures if the geothermal liquid is saline. *Anhydrite* is from the Greek *an* (= without) and *hydor* (= water).

210

Large gypsum crystals. From *Krafla, Northeast Iceland*. Size: *10 x 20 cm.*

From *Krýsuvík, Southwest Iceland.* Size: *9.3 x 6.7 cm.*

HALOTRICHITE

Crystal form: monoclinic
Hardness: 1½
Specific gravity: 1.89–1.95

Cleavage: not defined
$FeAl_2(SO_4)_4 \cdot 22H_2O$

DESCRIPTION: Halotrichite forms regular white or yellowish clusters with silky lustre, or fibrous tufts. It is soluble in water, and is hence best seen in dry conditions. It has a bitter taste.

OCCURRENCE: Halotrichite is precipitated from geothermal gases in patches of steaming ground in high-temperature areas. It is best collected after dry weather, as halotrichite is water-soluble and is washed away by rain. It was formerly used as a wood preservative.

NAME: *Halotrichite*, coined in 1839, is from the Greek *hals thrix* (= salty hair).

VARIANTS: *Pickeringite* is a magnesium variant of halotrichite. It is yellow or brown, but otherwise resembles halotrichite. It is water-soluble like halotrichite, and has the same bitter taste. *Pickeringite* was named in 1844 after Dedie A. J. *Pickering*.

From *Krýsuvík, Southwest Iceland.* Size: *11 x 8 cm.*

White halotrichite, yellow pickeringite and green brochantite. From *Krýsuvík, Southwest Iceland.* Size: *9.5 x 7 cm.*

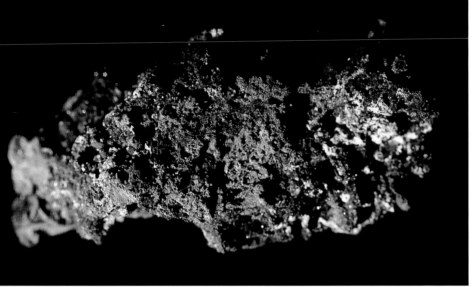

From *Krýsuvík, Southwest Iceland.* Size: *2 x 2 cm.*

BROCHANTITE

Crystal form: monoclinic
Hardness: 3½–4
Specific gravity: 3.9–4.0

Cleavage: perfect in one direction
$Cu_4SO_4(OH)_6$

DESCRIPTION: Brochantite forms acicular rods which grow in clusters, or coatings and crusts in which individual crystals are hardly discernible. Colour is green, and streak is pale green. Lustre is vitreous, or slightly pearly, on cleavage surfaces. It is translucent.

OCCURRENCE: In Iceland brochantite has only been found in geothermal areas, e.g. Reykjanes and Krýsuvík in the southwest and Fremrinámur in the north. When found in Krýsuvík in the 19th century it was thought to be a new mineral, and named *krisuvigite*.

NAME: *Brochantite* was named in 1824 after French mineralogist J.M *Brochant* de Villiers.

Brochantite and gypsum. From *Krýsuvík, Southwest Iceland*. Size: *7.3 x 4.5 cm*.

From *Krýsuvík, Southwest Iceland.* Size: 5 x 4 cm.

CHALCANTHITE

Crystal form: triclinic
Hardness: 2½
Specific gravity: 2.29

Cleavage: perfect
$CuSO_4 \cdot 5H_2O$

DESCRIPTION: Chalcanthite is bluish, blue-green or green. It forms short prismatic or massive tabular crystals, and sometimes fibrous spherules. Streak is white or colourless, and lustre is vitreous or greasy.

OCCURRENCE: Chalcanthite occurs in Iceland in patches of hot ground in high-temperature geothermal areas. Since it is water-soluble, it dissolves in rain, and is best seen after a period of dry weather.

NAME: *Chalcanthite* is from the Greek *chalkos* (= copper) and *anthos* (= flower).

From *Krýsuvík, Southwest Iceland*. Size: *4 x 3 cm.*

ALUNOGEN

Crystal form: trigonal
Hardness: 1½–2½
Specific gravity: 1.7

Cleavage: perfect in one direction if discernible

$Al_2(SO_4)_3 \cdot (OH)_6$

DESCRIPTION: Alunogen is colourless or white, sometimes greyish or yellow, and translucent. It forms friable sparse incrustations, up to 1 cm in thickness, consisting of acicular crystals (hair-salt). Alunogen has a bitter taste. It is water-soluble, so it is mainly observed in dry weather conditions.

OCCURRENCE: Alunogen occurs around fumaroles in geothermal areas, with other geothermal sublimates and sulphur.

NAME: *Alunogen* is from the Latin *alumen* (= alunite) and the Greek *genos* (= origin).

From *Skarðsheiði, West Iceland*. Size: *7.2 x 4.7 cm.*

JAROSITE

Crystal form: hexagonal
Hardness: 2½–3½
Specific gravity: 2.9–3.3

Cleavage: perfect in one direction
$KFe_3(SO_4)_2(OH)_6$

DESCRIPTION: Jarosite is yellow or brownish. It is very small-scale and forms thin flakes, fibrous coatings and small lumps. Streak is yellow. Jarosite is translucent with slight vitreous or greasy lustre. Jarosite is not easily distinguished from sulphur or brown ochre. Jarosite is a member of the alunite group of hydrous sulphates.

OCCURRENCE: Jarosite occurs in altered rhyolite, mainly on cleavage or fracture surfaces adjacent to hot springs or altered ground, e.g. around Torfajökull in the southern highlands, where it occurs on fractured rock walls and is often mistaken for sulphur.

NAME: *Jarosite* was named in 1852 after the place where it was first found, *Barenco Jaroso* in Spain.

VARIANTS: *Alunite* is a K(Na)Al sulphate which belongs to the same group of minerals as jarosite. It is white, but sometimes coloured by impurities. It occurs in volcanic areas where volcanic gases alter the rock.

From *Ölfus, South Iceland*. Size: *8 x 11 cm.*

HEMATITE (Solfataric hematite)

Crystal form: trigonal
Hardness: 5–6
Specific gravity: 5.3

Cleavage: none
Fe_2O_3

DESCRIPTION: Hematite is commonly deposited around fumaroles and solfataras in high-temperature geothermal areas. It forms irregular cavernous masses, up to 50 cm in thickness and rusty-brown in colour. While it bears a superficial resemblance to scoria, hematite is much heavier.

OCCURRENCE: Solfataric hematite is common in high-temperature geothermal areas. It is found as diffuse lumps and thick crusts on the surface, in patches of clayey alteration around solfataras. Being insoluble in water, it is also common as a residue in steam areas of declining activity.

MINERAL	CHEMICAL FORMULA	CRYSTAL SYSTEM	HARDNESS	
Actinolite	$Ca_2(Mg,Fe)_5Si_8O_{22}(OH,F)_2$	monoclinic	5½–6	
Agate/onyx	SiO_2	cryptocrystalline	7	
Alunite	$K(Na)Al_3(SO_4)_2(OH)_6$	hexagonal	3½–4	
Alunogen	$Al_2(SO_4)_3 \cdot 17H_2O$	trigonal	1½–2½	
Amethyst	SiO_2	hexagonal	7	
Analcime	$NaAlSi_2O_6 \cdot H_2O$	cubic	5–5½	
Apophyllite	$KFCa_4Si_8O_{20} \cdot 8H_2O$	tetragonal	4½–5	
Aragonite	$CaCO_3$	orthorhombic	3½–4	
Baryte	$BaSO_4$	orthorhombic	3–3½	
Biotite	$K(Mg,Fe)_3(Al,Fe)Si_3O_{10}(OH,F)_2$	monoclinic	2½–3	
Brochantite	$Cu_4SO_4(OH)_6$	monoclinic	3½–4	
Calcite	$CaCO_3$	trigonal	3	
Celadonite (illite)	$K(Mg,Fe)(Fe,Al)[(OH)_2 \,|\, Si_4O_{10}]$	monoclinic	2	
Chabazite	$CaAl_2Si_4O_{12} \cdot 6H_2O$	triclinic	4½	
Chalcanthite	$CuSO_4 \cdot 5H_2O$	triclinic	2½	
Chalcedony	SiO_2	cryptocrystalline	7	
Chalcopyrite	$CuFeS_2$	tetragonal	3½–4	
Chlorite	$(Fe,Mg,Al)_6(Si,Al)_4O_{10}(OH)_8$	monoclinic	2–3	
Citrine	SiO_2	hexagonal	7	
Covellite	CuS	hexagonal	1½–2	
Cowlesite	$CaAl_2Si_2O_8 \cdot 4H_2O$	monoclinic	2	
Cristobalite	SiO_2	tetragonal, cubic	6½–7	
Dolomite	$CaMg(CO_3)_2$	trigonal	3½–4	
Epidote	$Ca_2(Fe,Al)Al_2Si_3O_{12}(OH)$	monoclinic	6–7	
Epistilbite	$CaAl_2Si_6O_{16} \cdot 5H_2O$	monoclinic	4–4½	
Erionite	$(K_2,Ca,Na_2)_2Al_4Si_{14}O_{36} \cdot 15H_2O$	hexagonal	4	
Feldspar	$NaAlSi_3O_8 - CaAl_2Si_2O_8$ & $KAlSi_3O_8$	monoclinic, triclinic	6–6½	
Fluorite	CaF_2	cubic	4	
Galena	PbS	cubic	2½–3	
Garnet	$Ca_3Al_2[SiO_4]_3$	cubic	6½–7½	
Garronite	$Na_2Ca_5Al_{12}Si_{20}O_{64} \cdot 27H_2O$	orthorhombic	4½	
Gismondine	$Ca_2Al_4Si_4O_{16} \cdot 9H_2O$	monoclinic	4½	
Gypsum	$CaSO_4 \cdot 2H_2O$	monoclinic	2	
Gyrolite	$NaCa_{16}(Si_{23},Al)O_{60}(OH)_5 \cdot 15H_2O$	trigonal	3–4	
Halite	$NaCl$	cubic	2	
Halotrichite	$FeAl_2(SO_4)_4 \cdot 22H_2O$	monoclinic	1½	
Hedenbergite	$Ca(Fe,Mg)Si_2O_6$	monoclinic	5½–6	

Main colour / other colours	Streak	Cleavage	Lustre	Specific Gravity	Page
various greens, white, grey	white	perfect	vitreous, silky	2.9–3.3	196–197
white, grey, pale blue, yellowish, greenish	white	none	vitreous, greasy	2.57–2.65	147–149
white, greyish, yellowish, reddish, brownish	white	distinct	vitreous, pearly	2.6–2.7	218
colourless, white, greyish, yellow	white	perfect	vitreous, silky	1.7	217
violet	white	none	vitreous, greasy	2.65	140–141
colourless, white / grey, pink, yellow	white	indistinct	vitreous	2.22–2.63	122–123
white, colourless / greenish, yellowish, reddish	white	perfect	vitreous, pearly	2.33–2.37	124–125
colourless, greyish / white, yellowish, bluish, reddish	white	distinct	vitreous, pearly	2.95	162–163
white, greyish, yellowish	white	perfect	vitreous	4.5	170–171
dark brown, black	white	perfect	pearly	2.7–3.3	38
green	pale green	perfect	vitreous, pearly	3.9–4.0	214–215
colourless, white, other pale colours / yellowish, yellow-brown	white	perfect	vitreous	2.7	155-157, 207
blue-green, green, blue		perfect		2.95–3.05	204
white, colourless / reddish, yellowish, brownish	white	distinct	vitreous	1.97–2.20	120–121
blue, blue-green, green	white, colourless	perfect	vitreous, greasy	2.29	216
white, grey, pale blue, yellowish, greenish	white	none	vitreous, greasy	2.57–2.65	144–146
yellowish, multicoloured	black, greenish	indistinct	metallic	4.2	182–183
green: from very pale to near-black	green	perfect	vitreous, pearly	about 2.8	202–203
yellowish	white	none	vitreous, greasy	2.65	139
blue, violet, pale blue, dark blue	dark blue	perfect	metallic	4.7	186
white, colourless, grey	white	perfect	vitreous, pearly	2.05–2.14	106–107
white	white	none		2.2	143
colourless, white / yellow, brown, grey, black	white	perfect	vitreous, pearly	2.86	164–165
dark green to yellow-green / grey, black	white to greyish	perfect	vitreous	3.3–3.5	190–191
colourless, white, yellowish, bluish	white	perfect	vitreous, pearly	2.22–2.68	112–113
white	white	indistinct	pearly	2.02–2.13	118–119
white colourless, yellowish, grey, greenish, reddish	white	perfect	vitreous, pearly	2.61–2.76	32–33
colourless, greenish, violet / any colour	white	perfect	vitreous	3.1 – 3.2	168–169
leaden grey	leaden-grey to black	perfect	metallic	7.5	185
brownish, reddish-brown	white	indistinct	vitreous	3.4–3.6	194–195
white, colourless	white	perfect	vitreous	2.13–2.18	104–105
white, colourless	white	indistinct	vitreous	2.12–2.28	102–103
white, colourless / grey, yellow, red, brown	white	perfect	vitreous, pearly	2.3–2.4	210–211
white	white	perfect	silky, vitreous	2.3	126–127
white, grey, colourless / yellow, reddish, bluish	white	perfect	vitreous	2.1–2.2	172
white, yellowish	white		vitreous, silky	1.89–1.95	212–213
dark green to black	white to pale green	perfect	vitreous	3.2–3.6	198

Mineral	Chemical formula	Crystal system	Hardness
Hematite	Fe_2O_3	trigonal	5–6
Heulandite	$(Na,Ca)_{2-3}Al_3(Al,Si)_2Si_{13}O_{36}\cdot12H_2O$	monoclinic	$3\frac{1}{2}$–4
Hornblende	$Ca_2(Mg,Fe)_4Al(Si_7Al)O_{22}(OH,F)$	monoclinic	5–6
Iceland spar	$CaCO_3$	trigonal	3
Ilvaite	$CaFe_3OSi_2O_7(OH)$	orthorhombic	$5\frac{1}{2}$–6
Jarosite	$KFe_3(SO_4)_2(OH)_6$	hexagonal	$2\frac{1}{2}$–$3\frac{1}{2}$
Jasper	SiO_2	cryptocrystalline	7
Kaolinite	$Al_2Si_2O_5(OH)_4$	triclinic	2–$2\frac{1}{2}$
Laumontite	$CaAl_2Si_4O_{12}\cdot4H_2O$	monoclinic	3–$3\frac{1}{2}$
Levyne	$(Ca,Na_2,K_2)Al_2Si_4O_{12}\cdot6H_2O$	hexagonal	4–$4\frac{1}{2}$
Limonite	$FeOOH\cdot nH_2O$	near-amorphous	4–$5\frac{1}{2}$
Magnetite	Fe_3O_4	cubic	$5\frac{1}{2}$–6
Malachite	$Cu_2(OH)_2CO_3$	monoclinic	$3\frac{1}{2}$–4
Mesolite	$Na_2Ca_2Al_6Si_9O_{30}\cdot8H_2O$	monoclinic	5
Mordenite	$(Na_2K_2Ca)Al_2Si_{10}O_{24}\cdot7H_2O$	orthorhombic	4–5
Natrolite	$Na_2Al_2Si_3O_{10}\cdot2H_2O$	orthorhombic	5–$5\frac{1}{2}$
Okenite	$Ca_{10}Si_{18}O_{46}\cdot18H_2O$	triclinic	$4\frac{1}{2}$–5
Olivine	$(Mg,Fe)_2SiO_4$	orthorhombic	$6\frac{1}{2}$–7
Opal	$SiO_2\cdot nH_2O$	amorphous	$5\frac{1}{2}$–$6\frac{1}{2}$
Phillipsite	$(K,Na,Ca)_{1-2}(Si,Al)_8O_{16}\cdot6H_2O$	monoclinic	$4\frac{1}{2}$
Pickeringite	$MgAl_2(SO_4)_4\cdot22H_2O$	monoclinic	$1\frac{1}{2}$
Prehnite	$Ca_2Al_2Si_3O_{10}(OH)_2$	orthorhombic	6–$6\frac{1}{2}$
Pyrite	FeS_2	cubic	6–$6\frac{1}{2}$
Pyroxene	$(Ca,Mg,Al,Ti)_2(Si,Al)_2O_6$	monoclinic	$5\frac{1}{2}$–6
Quartz	SiO_2	hexagonal	7
Reyerite	$(Na,K)_2Ca_{14}(Si,Al)_{24}O_{58}(OH)_8\cdot6H_2O$	hexagonal	$3\frac{1}{2}$–$4\frac{1}{2}$
Rosasite	$(Cu,Zn)_2CO_3(OH)_2$	monoclinic	4–$4\frac{1}{2}$
Scolecite	$CaAl_2Si_3O_{10}\cdot3H_2O$	monoclinic	5
Siderite	$FeCO_3$	trigonal	4–$4\frac{1}{2}$
Smectite (montmorillonite)	$(Na,Ca)(Al,Mg)_6(Si_4O_{10})_3(OH)_6\cdot nH_2O$	monoclinic	1–2
Smoky quartz	SiO_2	hexagonal	7
Sphalerite	ZnS	cubic	$3\frac{1}{2}$–4
Stilbite	$NaCa_2Al_5Si_{13}O_{36}\cdot14H_2O$	monoclinic, orthorhombic	$3\frac{1}{2}$–4
Sugar calcite	$CaCO_3$	trigonal	3
Sulphur	S	orthorhombic	2
Thaumasite	$Ca_3Si(CO_3)(SO_4)(OH)_6\cdot12H_2O$	trigonal	$3\frac{1}{2}$
Thomsonite	$NaCa_2Al_5Si_5O_{10}\cdot6H_2O$	orthorhombic	5
Yugawaralite	$CaAl_2Si_6O_{16}\cdot4H_2O$	monoclinic	$4\frac{1}{2}$

Main colour / other colours	Streak	Cleavage	Lustre	Specific Gravity	Page
steel-grey, blackish, red, red-brown	reddish-brown, black	none	metallic	5.3	176-177, 219
colourless / white, grey, yellow, reddish	white	perfect	pearly, vitreous	2.1–2.29	110–111
black / dark green, green	white, grey	perfect	vitreous	3.2	39
colourless	white	perfect	vitreous	2.7	158–159
black, greyish-black	black, brownish	none	vitreous, dull metallic	3.8–4.1	132–133
pale yellow, brownish, brown	pale yellow	distinct	vitreous, greasy	2.9–3.3	218
many colours	white	none	vitreous, greasy	2.57–2.65	150–151
white, grey, pale blue / reddish, brownish, bluish	white	perfect	none	2.6	200
white / colourless, pale pink, reddish-brown	white	perfect	vitreous, pearly	2.20–2.41	98–99
colourless, white	white	indistinct	vitreous	2.09–2.16	114–115
dark brown, yellow-brown / red-brown, yellow	brown	indistinct	none/dull metallic	2.7–4.3	174–175
black	black	indistinct	metallic	5.2	178–179
pale to dark green	pale green	distinct	vitreous, silky	4	187
white, colourless, grey, pink, reddish	white	perfect	vitreous, silky	2.25–2.26	92–93
white / yellowish, reddish	white	perfect	vitreous, silky	2.12–2.15	94–95
colourless, white / grey, yellowish, brownish, reddish	white	perfect	vitreous, silky	2.20–2.26	92
white, colourless	white	perfect	vitreous, pearly	2.3	128–129
olive green, yellow-green, brown / colourless	white	indistinct	vitreous	3.2–4.3	36–37
colourless, any colour	white	none	vitreous	1.9–2.3	152-153, 206
colourless, white, pale brown / yellowish, grey	white	distinct	vitreous	2.2	100–101
yellow, brown, white		indistinct	vitreous, silky	1.79–1.90	212
pale green / white, greyish, yellow-green	white	distinct	vitreous	2.9	192–193
yellow, yellow-brown	black	none	metallic	4.9–5.1	180–181
black to dark green	grey to green	perfect	vitreous	3.4	34–35
white, colourless, grey, various colours	white	none	vitreous, greasy	2.65	135
white, colourless	white	perfect		2.54–2.58	131
blue-green, green, blue	pale green	perfect	vitreous, none	4.0–4.2	188
white, colourless / grey, yellowish, brownish, reddish	white	perfect	vitreous, silky	2.25–2.31	90-91
yellow, brown, red-brown, white, grey, black	white to pale brown	perfect	vitreous	3.7–3.9	166–167
brownish, greenish / black		perfect	none	2.5	201
light brown, brown	white	none	vitreous, greasy	2.65	142
yellow, brown, dark grey to black	yellow-brown	perfect	metallic	3.9–4.1	184
white, colourless / yellow, reddish. grey, brownish	white	perfect	vitreous, pearly	2.12–2.21	108–109
yellow-brown	white	perfect	vitreous	2.7	160–161
yellow, pale yellow / reddish	yellow to white	indistinct	greasy	2	208–209
white, colourless	white	indistinct	vitreous, silky	1.9	130
white, grey, pale blue, reddish, yellowish	white	perfect	vitreous, pearly	2.25–2.44	96–97
colourless, white	white	distinct	vitreous	2.19–2.25	116–117

PUBLICATIONS ON ICELANDIC MINERALS AND ROCKS

BOOKS

Ari Trausti Guðmundsson and Halldór Kjartansson. *Earth in Action.* 166 pp. Vaka-Helgafell 1996.

Axel Kaaber, Einar Gunnlaugsson and Kristján Sæmundsson. *Íslenskir steinar.* 1st. ed. 143 pp. Bókaútgáfan Bjallan 1988.

Guðmundur Páll Ólafsson. *Iceland the Enchanted* (translation of *Perlur í náttúru Íslands*). 420 pp. Mál og menning 1995.

Hjálmar R. Bárðarson. *Íslenskt grjót.* 1st ed. 288 pp. Hjálmar R. Bárðarson 1995.

Þorleifur Einarsson. *Geology of Iceland. Rocks and Landscape* (Translation of *Jarðfræði. Saga bergs og lands*) 1st. ed. 309 pp. Mál og menning 1994.

ARTICLES

Betz, Volker. Zeolites from Iceland and the Faeroes. *The Mineralogical Record* 1-2, 1981, pp. 5-26.

Halldór Kjartansson. Holufyllingar. *Útivist* vol. 5, 1979, pp. 33-57.

Haraldur Sigurðsson. Geology of the Setberg area, Snæfellsnes, western Iceland. *Greinar* IV, 2., *Vísindafélag Íslendinga* 1966, pp. 53-125.

Lapis. German-language amateur periodical on rocks. Issues on zeolites and Icelandic rocks vol. 3, no. 1 and 10, 1978.

Steinn. Amateur periodical on petrology.

Sveinn Jakobsson. Íslenskar bergtegundir I. – Pikrít (óseanít) [Icelandic rock types I Picrite (oceanite)]. *Náttúrufræðingurinn* vol. 52, 1983, pp. 80-85.

Sveinn Jakobsson. Íslenskar bergtegundir II. – Ólivínþóleiít [Icelandic rock types II Olivine tholeiite]. *Náttúrufræðingurinn* vol. 53, 1984, pp. 13-18.

Sveinn Jakobsson. Íslenskar bergtegundir III. – Þóleiít [Icelandic rock types III Tholeiite]. *Náttúrufræðingurinn* vol. 53, 1984, pp. 53-59.

Sveinn Jakobsson. Íslenskar bergtegundir IV. – Basaltískt íslandít og íslandít [Icelandic rock types IV Basaltic icelandite and icelandite]. *Náttúrufræðingurinn* vol. 54, 1985, pp. 77-84.

Sveinn Jakobsson. Íslenskar bergtegundir V. – Dasít (rýódasít) [Icelandic rock types V Dacite (rhyodacite)]. *Náttúrufræðingurinn* vol. 54, 1985, pp. 149-153.

Sveinn Jakobsson. Íslenskir zeólítar (geislasteinar). [Icelandic zeolites]. *Árbók Ferðafélags Íslands* 1977, pp. 190-201.

Sveinn Jakobsson. Outline of the petrology of Iceland. *Jökull* vol. 29, 1979, pp. 57-73.

Walker, George P.L. Geology of the Reyðarfjörður area, eastern Iceland. *Quart. Jour. Geol. Soc. London* vol. 114, 1959, pp. 367-393.

Walker, George P.L. The Breiðdalur central volcano, eastern Iceland. *Quart. Jour. Geol. Soc. London* vol. 119, 1963, pp. 29-63.

Walker, George P.L. The structure of eastern Iceland. In *Geodynamics of Iceland and the North Atlantic Area* (Ed. Leó Kristjánsson) 1974, pp. 177-188.

Walker, George P.L. Zeolite zones and dike distribution in relation to the structure of the basalts of eastern Iceland. *The Journal of Geology* vol. 68, 1960, pp. 515-528.

OWNERS OF ILLUSTRATED ROCKS

P. 17 Mesolite-apophyllite-gismondine. Owner Hermann Tönsberg.
P. 18 upper. Stilbite fissure filling. Owner Kristján Sæmundsson.
P. 18 lower. Chalcedony-heulandite-sugar calcite. Owner Kristján Sæmundsson.
P. 32 Plagioclase phenocryst in basalt. Owner Kristján Sæmundsson.
P. 33 upper. Plagioclase crystals. Owner Hermann Tönsberg.
P. 33 lower. Potassium feldspar. Owner Hermann Tönsberg.
P. 34 Pyroxene in gabbro. Owner Icelandic Museum of Natural History (Axel Kaaber collection).
P. 35 upper. Pyroxene phenocrysts in hyaloclastite. Owner Steinaríki Íslands (Mineral Kingdom, Akranes).
P. 35 lower. Pyroxene crystals. Owner Hermann Tönsberg.
P. 36 Olivine phenocrysts in picrite. Owner Kristján Sæmundsson.
P. 37 upper. Olivine crystals. Owner Hermann Tönsberg.
P. 37 lower. Olivine altered into iddingsite. Owner Hermann Tönsberg.
P. 38 Biotite. Owner Icelandic Museum of Natural History (Acq. no. 9600).
P. 39 Hornblende. Owner Icelandic Museum of Natural History (Acq. no. 15532).
P. 42 Gabbro. Owner Einar Gunnlaugsson.
P. 43 upper. Gabbro xenolith. Owner Kristján Sæmundsson.
P. 43 lower. Anorthosite. Owner Hrefna Kristmannsdóttir.
P. 44 Dolerite. Owner Kristján Sæmundsson.
P. 45 Diorite. Owner University of Iceland.
P. 46 Microgranite. Owner Einar Gunnlaugsson.
P. 47 Granophyre. Owner Kristján Sæmundsson.
P. 48 Gabbro/granophyre composite rock. Owner Kristján Sæmundsson.
P. 50 Olivine basalt. Owner Steinaríki Íslands (Mineral Kingdom, Akranes).
P. 51 upper. Olivine basalt with pyroxene crystals. Owner Steinaríki Íslands (Mineral Kingdom, Akranes).
P. 51 lower. Altered olivine basalt. Owner Kristján Sæmundsson.
P. 52 Tholeiite. Owner Kristján Sæmundsson.
P. 53 Porphyritic basalt. Owner Kristján Sæmundsson.
P. 54 Scoria. Owner Hermann Tönsberg.
P. 55 upper. Scoria. Owner Kristján Sæmundsson.
P. 55 lower left. Black lapilli. Owner Einar Gunnlaugsson.
P. 55 lower right. Pele's hair. Owner Einar Gunnlaugsson.
P. 56 Palagonite tuff. Owner Kristján Sæmundsson.
P. 57 lower. Palagonite. Owner Kristján Sæmundsson.
P. 58 upper. Breccia. Owner Kristján Sæmundsson.
P. 58 lower left. Altered hyaloclastite. Owner Kristján Sæmundsson.
P. 58 lower right. Tachylite. Owner Kristján Sæmundsson.
P. 59 Picrite. Owner Kristján Sæmundsson.
P. 60 Ankaramite. Owner Einar Gunnlaugsson.
P. 61 Basalt/rhyolite composite rock. Owner Kristján Sæmundsson.
P. 62 Andesite. Owner Kristján Sæmundsson.
P. 63 upper. Dacite. Owner Kristján Sæmundsson.

P. 63 lower. Andesite. Owner Kristján Sæmundsson.

P. 64 Flow-banded rhyolite. Owner Kristján Sæmundsson.

P. 65 Rhyolite. Owner Steinaríki Íslands (Mineral Kingdom, Akranes).

P. 66 Pumice. Owner Kristján Sæmundsson.

P. 67 Obsidian. Owner Einar Gunnlaugsson.

P. 68 Pitchstone. Owner Hermann Tönsberg.

P. 69 upper. Porphyritic pitchstone. Owner Kristján Sæmundsson.

P. 69 lower. Acidic breccia. Owner Kristján Sæmundsson.

P. 70 Perlite. Owner Hlíðaskóli.

P. 71 upper. Perlite. Owner University of Iceland.

P. 71 lower. Crumbled perlite. Owner Kristján Sæmundsson.

P. 72 Volcanic bomb with xenolith. Owner Kristján Sæmundsson.

P. 73 Gabbro xenolith in basalt. Owner Kristján Sæmundsson.

P. 74 Breccia. Owner Kristján Sæmundsson.

P. 75 Breccia. Owner Maggý Valdimarsdóttir.

P. 76 Spherulites. Owner Grétar Eiríksson.

P. 77 upper left. Spherulites. Owner Maggý Valdimarsdóttir.

P. 77 upper right. Spherulites. Owner Icelandic Museum of Natural History (Axel Kaaber collection).

P. 77 lower. Spherulites. Owner Kristján Sæmundsson.

P. 79 Claystone. Owner Kristján Sæmundsson.

P. 80 Sandstone. Owner Kristján Sæmundsson.

P. 81 Interbasaltic sediments. Owner Kristján Sæmundsson.

P. 82 Fossil leaf. Owner Icelandic Museum of Natural History (Axel Kaaber collection).

P. 83 Fossil shells. Owner Kristján Sæmundsson.

P. 84 Lignite. Owner Icelandic Museum of Natural History (Acq. no. 423).

P. 85 upper left. Bituminous claystone. Owner Folk Museum, Höfn, Hornafjörður.

P. 85 upper right. Flaky petrified lignite. Owner Icelandic Museum of Natural History (Acq. no. 489).

P. 85 lower. Petrified wood. Owner Kristján Sæmundsson.

P. 86 Silicified wood. Owner Kristín Einarsdóttir.

P. 87 Crawl trails. Owner Kristján Sæmundsson.

P. 90 Scolecite. Owner Icelandic Museum of Natural History.

P. 91 upper left. Scolecite. Owner Einar Gunnlaugsson.

P. 91 upper right. Scolecite. Icelandic Museum of Natural History.

P. 91 lower. Scolecite. Owner Einar Gunnlaugsson.

P. 92 Mesolite. Owner family at Volasel, Lón.

P. 93 upper left. Mesolite. Owner Icelandic Museum of Natural History (Axel Kaaber collection).

P. 93 upper right. Mesolite. Owner Icelandic Museum of Natural History (Acq. no. 14495).

P. 93 lower. Mesolite. Owner Icelandic Museum of Natural History (Acq. no. 6967).

P. 94 Mordenite. Axel Kaaber collection.

P. 95 upper. Mordenite. Owner Icelandic Museum of Natural History (Acq. no. 3928).

P. 95 lower. Mordenite. Owner Hermann Tönsberg.

P. 96 Thomsonite. Owner Icelandic Museum of Natural History (Acq. no. 14267).

P. 97 upper left. Thomsonite. Owner Hermann Tönsberg.

P. 97 upper right. Thomsonite. Owner Kristján Sæmundsson.

P. 97 lower left. Thomsonite. Owner Hermann Tönsberg.

P. 97 lower right. Thomsonite. Owner Steinaríki Íslands (Mineral Kingdom, Akranes).
P. 98 Laumontite. Owner Steinaríki Íslands (Mineral Kingdom, Akranes).
P. 99 upper. Laumontite. Owner Svavar Guðmundsson.
P. 99 lower. Laumontite. Owner Hermann Tönsberg.
P. 100 Phillipsite. Owner Hermann Tönsberg.
P. 101 upper left. Phillipsite. Owner Hermann Tönsberg.
P. 101 upper right. Phillipsite. Owner Hermann Tönsberg.
P. 101 lower left. Phillipsite. Owner Icelandic Museum of Natural History.
P. 101 lower right. Phillipsite. Owner Steinaríki Íslands (Mineral Kingdom, Akranes).
P. 102 Gismondine. Owner Icelandic Museum of Natural History (Acq. no. 15416).
P. 103 upper. Gismondine. Owner Icelandic Museum of Natural History (Acq. no. 10757).
P. 103 lower left. Gismondine. Owner Sigurður S. Jónsson.
P. 103 lower right. Gismondine. Owner Hermann Tönsberg.
P. 104 Garronite. Owner Steinaríki Íslands (Mineral Kingdom, Akranes).
P. 105 upper. Garronite. Owner Icelandic Museum of Natural History (Acq. no. 11935).
P. 105 lower Garronite. Owner Steinaríki Íslands (Mineral Kingdom, Akranes).
P. 106 Cowlesite. Owner Hermann Tönsberg.
P. 107 upper. Cowlesite. Owner Sigurður S. Jónsson.
P. 107 lower. Cowlesite. Owner Hermann Tönsberg.
P. 108 Stilbite. Owner Kristján Sæmundsson.
P. 109 upper left. Stilbite. Owner Folk Museum, Höfn, Hornafjörður.
P. 109 upper right. Stilbite. Owner Kristján Sæmundsson.
P. 109 middle left. Stilbite. Owner Kristján Sæmundsson.
P. 109 middle right. Stilbite. Owner Gunnhildur Á. Steingrímsdóttir.
P. 109 lower left. Stilbite. Owner Kristján Sæmundsson.
P. 109 lower right. Stilbite. Axel Kaaber collection.
P. 110 Heulandite. Owner Icelandic Museum of Natural History (Acq. no. 557).
P. 111 upper left. Heulandite. Owner Björgólfur Jónsson.
P. 111 upper right. Heulandite. Owner Einar Gunnlaugsson.
P. 111 lower. Heulandite. Owner Steinaríki Íslands (Mineral Kingdom, Akranes).
P. 112 Epistilbite. Owner Hermann Tönsberg.
P. 113 upper left. Epistilbite. Owner Hermann Tönsberg.
P. 113 upper right. Epistilbite. Owner Hermann Tönsberg.
P. 113 lower. Epistilbite. Owner Hermann Tönsberg.
P. 114 Levyne. Owner Icelandic Museum of Natural History (Acq. no. 7932).
P. 115 upper left. Levyne. Owner Hermann Tönsberg.
P. 115 upper right. Levyne. Owner Sigurður S. Jónsson.
P. 115 lower. Levyne. Owner Steinaríki Íslands (Mineral Kingdom, Akranes).
P. 116 Yugawaralite. Owner Sigurður S. Jónsson.
P. 117 upper. Yugawaralite. Owner Sigurður S. Jónsson.
P. 117 lower. Yugawaralite. Owner Hermann Tönsberg.
P. 118 Erionite. Owner Kristján Sæmundsson.
P. 119 Erionite. Owner Kristján Sæmundsson.
P. 120 Chabazite. Owner Kristján Sæmundsson.
P. 121 upper left. Chabazite. Owner Kristján Sæmundsson.
P. 121 upper right. Chabazite. Owner Steinaríki Íslands (Mineral Kingdom, Akranes).
P. 121 lower left. Chabazite. Owner Steinaríki Íslands (Mineral Kingdom, Akranes).

P. 121 lower right. Chabazite. Owner Hermann Tönsberg.
P. 122 Analcime.
P. 123 upper. Analcime. Owner Hermann Tönsberg.
P. 123 lower left. Wairakite. Owner Steinaríki Íslands (Mineral Kingdom, Akranes).
P. 123 lower right. Analcime. Owner Kristján Sæmundsson.
P. 124 Apophyllite. Owner Steinaríki Íslands (Mineral Kingdom, Akranes).
P. 125 upper. Apophyllite. Owner Hildigunnur Þorsteinsdóttir.
P. 125 lower left. Apophyllite. Owner Einar Gunnlaugsson.
P. 125 lower right. Apophyllite. Owner Hermann Tönsberg.
P. 126 Gyrolite. Owner Icelandic Museum of Natural History (Acq. no. 6372).
P. 127 upper. Gyrolite. Owner Icelandic Museum of Natural History (Acq. no. 14103).
P. 127 lower. Gyrolite. Owner Steinaríki Íslands (Mineral Kingdom, Akranes).
P. 128 Okenite. Owner Icelandic Museum of Natural History (Acq. no. 15498).
P. 129 upper. Okenite. Owner Steinaríki Íslands (Mineral Kingdom, Akranes).
P. 129 lower. Okenite. Owner Hermann Tönsberg.
P. 130 Thaumasite. Owner Hermann Tönsberg.
P. 131 Reyerite. Owner Hermann Tönsberg.
P. 132 Ilvaite. Owner Hermann Tönsberg.
P. 133 upper. Ilvaite. Owner Sigurður S. Jónsson.
P. 133 lower. Ilvaite. Owner Hermann Tönsberg.
P. 135 Quartz. Owner Kristján Sæmundsson.
P. 136 Rock crystal. Owner family at Volasel, Lón.
P. 137 upper. Rock crystal. Owner Grétar Eiríksson.
P. 137 lower left. Rock crystal. Owner Sigurður S. Jónsson.
P. 137 lower right. Rock crystal. Owner Björgólfur Jónsson.
P. 138 Rock crystal. Owner Icelandic Museum of Natural History.
P. 139 Citrine. Owner Hermann Tönsberg.
P. 140 Amethyst and Calcite. Owner Álfasteinn, Borgarfjörður eystri.
P. 141 Amethyst. Owner Friðrik Jónsson, Hraunkot, Lón.
P. 142 Smoky quartz. Owner Hermann Tönsberg.
P. 143 Cristobalite. Owner Steinaríki Íslands (Mineral Kingdom, Akranes).
P. 144 Chalcedony. Owner Icelandic Museum of Natural History (Acq. no. 455).
P. 145 upper. Chalcedony. Owners Sigurborg Í. Einarsdóttir and Sören Sörensen, Eskifjörður.
P. 145 lower left. Chalcedony. Owner Kristján Sæmundsson.
P. 145 lower right. Chalcedony. Owner Einar Gunnlaugsson.
P. 146 upper. Chalcedony. Owner Gísli Arason, Höfn, Hornafjörður.
P. 146 lower. Chalcedony. Owners Sigurborg Í. Einarsdóttir and Sören Sörensen, Eskifjörður.
P. 147 Onyx. Owner Friðrik Jónsson, Hraunkot, Lón.
P. 148 Agate. Owner Petra Sveinsdóttir, Stöðvarfjörður.
P. 149 upper. Agate. Owners Sigurborg Í. Einarsdóttir and Sören Sörensen, Eskifjörður.
P. 149 lower. Moss agate. Owners Sigurborg Í. Einarsdóttir and Sören Sörensen, Eskifjörður.
P. 150 Jasper. Owner Einar Gunnlaugsson.
P. 151 upper left. Jasper. Owner Grétar Eiríksson.
P. 151 upper right. Jasper. Owner Steinaríki Íslands (Mineral Kingdom, Akranes).
P. 151 lower. Jasper. Owner Grétar Eiríksson.
P. 152 Opal. Owner Icelandic Museum of Natural History (Acq. no. 7179).
P. 153 upper left. Opal. Owner Icelandic Museum of Natural History (Acq. no. 1304).

P. 153 upper right. Opal. Owner Hermann Tönsberg.
P. 153 lower. Opal. Owner Kristján Sæmundsson.
P. 155 Calcite. Owner Hermann Tönsberg.
P. 156 upper left. Calcite. Owner Sigurður S. Jónsson.
P. 156 upper right. Calcite. Owner Kristján Sæmundsson.
P. 156 lower. Calcite. Owner Heiðveig Guðlaugsdóttir, Hoffell, Hornafjörður.
P. 157 upper. Calcite. Owner Icelandic Museum of Natural History.
P. 157 lower left. Calcite. Owner Álfasteinn, Borgarfjörður eystri.
P. 157 lower right. Calcite. Axel Kaaber collection.
P. 158 Iceland spar. Owner Heiðveig Guðlaugsdóttir, Hoffell, Hornafjörður.
P. 159 upper. Iceland spar. Owner Geological Museum, London.
P. 159 lower. Iceland spar. Owner Icelandic Museum of Natural History (Acq. no. 501).
P. 160 Sugar calcite. Axel Kaaber collection.
P. 161 upper. Sugar calcite. Owner Álfasteinn, Borgarfjörður eystri.
P. 161 lower. Sugar calcite. Owner Gísli Arason, Höfn, Hornafjörður.
P. 162 Aragonite. Owner Hermann Tönsberg.
P. 163 upper. Aragonite. Owner Kristján Sæmundsson.
P. 163 lower left. Aragonite. Axel Kaaber collection.
P. 163 lower right. Flos ferri. Owner Svavar Guðmundsson.
P. 164 Dolomite. Owner Hermann Tönsberg.
P. 165 upper. Dolomite on quartz. Owner Hermann Tönsberg.
P. 165 lower. Dolomite. Owner Hermann Tönsberg.
P. 166 Siderite. Owner Hermann Tönsberg.
P. 167 upper. Siderite. Owner Hermann Tönsberg.
P. 167 lower left. Siderite. Owner Hermann Tönsberg.
P. 167 lower right. Siderite. Owner Hermann Tönsberg.
P. 168 Fluorite. Owner Björgólfur Jónsson, Tungufell, Breiðdalur.
P. 169 upper. Fluorite. Owner Álfasteinn, Borgarfjörður eystri.
P. 169 lower left. Fluorite. Owner Björgólfur Jónsson, Tungufell, Breiðdalur.
P. 169 lower right. Fluorite. Owner Steinaríki Íslands (Mineral Kingdom, Akranes).
P. 170 Baryte. Owner Icelandic Museum of Natural History.
P. 171 Baryte. Owner Björgólfur Jónsson, Tungufell, Breiðdalur.
P. 172 Rock salt. Owner Hermann Tönsberg.
P. 174 Limonite. Owner Hermann Tönsberg.
P. 175 top left. Limonite.
P. 175 top right. Limonite. Owner Hermann Tönsberg.
P. 175 centre. Limonite. Owner Steinaríki Íslands (Mineral Kingdom, Akranes).
P. 175 lower left. Limonite. Owner Kristján Sæmundsson.
P. 175 lower right. Limonite. Owner Álfasteinn, Borgarfjörður eystri.
P. 176 Hematite. Owner Hermann Tönsberg.
P. 177 upper left. Hematite. Owner Steinaríki Íslands (Mineral Kingdom, Akranes).
P. 177 upper right. Hematite. Owner Hermann Tönsberg.
P. 177 lower. Hematite. Owner Hermann Tönsberg.
P. 178 Magnetite in olivine basalt. Owner Steinaríki Íslands (Mineral Kingdom, Akranes).
P. 179 upper. Magnetite. Owner Sigurður S. Jónsson.
P. 179 lower. Magnetite. Owner Hermann Tönsberg.
P. 180 Pyrite. Owner Grétar Eiríksson.

P. 181 upper left. Pyrite. Axel Kaaber collection.
P. 181 upper right. Pyrite. Owner Icelandic Museum of Natural History.
P. 181 lower left. Pyrite. Owner Kristján Sæmundsson.
P. 181 lower right. Pyrite. Owner Steinaríki Íslands (Mineral Kingdom, Akranes).
P. 182 Chalcopyrite. Owner Hermann Tönsberg.
P. 183 upper. Chalcopyrite. Owner Hermann Tönsberg.
P. 183 lower. Chalcopyrite. Owner Hermann Tönsberg.
P. 184 Sphalerite. Owner Hermann Tönsberg.
P. 185 Galena. Owner Hermann Tönsberg.
P. 186 Covellite. Owner Hermann Tönsberg.
P. 187 Malachite. Owner Hermann Tönsberg.
P. 188 Rosasite. Owner Icelandic Museum of Natural History.
P. 190 Epidote. Owner Hermann Tönsberg.
P. 191 upper. Epidote. Owner Sigurður S. Jónsson.
P. 191 lower. Epiote. Owner Hermann Tönsberg.
P. 192 Prehnite. Owner Icelandic Museum of Natural History.
P. 193 Prehnite. Owner Icelandic Museum of Natural History.
P. 194 Garnet. Owner Hermann Tönsberg.
P. 195 upper. Garnet. Owner Hermann Tönsberg.
P. 195 lower. Garnet. Owner Hermann Tönsberg.
P. 196 Actinolite. Owner Einar Gunnlaugsson.
P. 197 upper. Actinolite. Owner Icelandic Museum of Natural History (Acq. no. 11659).
P. 197 lower. Actinolite. Owner Hermann Tönsberg.
P. 198 Hedenbergite. Owner Guðmundur Ómar Friðleifsson.
P. 200 Kaolinite. Owner Einar Gunnlaugsson.
P. 201 left. Smectite. Owner Einar Gunnlaugsson.
P. 201 right Smectite. Owner Einar Gunnlaugsson.
P. 202 Chlorite. Owner Hermann Tönsberg.
P. 203 upper. Chlorite. Owner Kristján Sæmundsson.
P. 203 lower left. Chlorophaeite. Owner Hermann Tönsberg.
P. 203 lower right. Chlorophaeite. Owner Hermann Tönsberg.
P. 204 Celadonite. Owner Einar Gunnlaugsson.
P. 206 Silica sinter. Owner Kristján Sæmundsson.
P. 207 Travertine. Owner Kristján Sæmundsson.
P. 208 Sulphur. Axel Kaaber collection.
P. 209 Sulphur. Owner Einar Gunnlaugsson.
P. 210 Gypsum. Owner Kristján Sæmundsson.
P. 211 Gypsum. Axel Kaaber collection.
P. 212 Halotrichite. Owner Kristján Sæmundsson.
P. 213 upper. Halotrichite. Owner Kristján Sæmundsson.
P. 213 lower. Halotrichite. Owner Kristján Sæmundsson.
P. 214 Brochantite. Owner Hermann Tönsberg.
P. 215 Brochantite and gypsum. Owner Steinaríki Íslands (Mineral Kingdom, Akranes).
P. 216 Chalcanthite. Owner Hermann Tönsberg.
P. 217 Alunogen. Owner Hermann Tönsberg.
P. 218 Jarosite. Owner Icelandic Museum of Natural History.
P. 219 Solfataric hematite. Owner Einar Gunnlaugsson.

INDEX